THE BEST AUSTRALIAN SCIENCE WRITING 2023

EDITOR
DONNA LU

FOREWORD
MARY O'KANE

NEWSOUTH

UNSW Press acknowledges the Bedegal people, the Traditional Owners of the unceded territory on which the Randwick and Kensington campuses of UNSW are situated, and recognises their continuing connection to Country and culture. We pay our respects to Bedegal Elders past and present.

A NewSouth book

Published by
NewSouth Publishing
University of New South Wales Press Ltd
University of New South Wales
Sydney NSW 2052
AUSTRALIA
https://unsw.press

 A catalogue record for this book is available from the National Library of Australia

ISBN   9781742238005 (paperback)
        9781742238883 (ebook)
        9781742239828 (ePDF)

Design Josephine Pajor-Markus
Cover design George Saad

All reasonable efforts were taken to obtain permission to use copyright material reproduced in this book, but in some cases copyright could not be traced. The editor welcomes information in this regard.

THE BEST AUSTRALIAN SCIENCE WRITING 2023

DONNA LU is *Guardian Australia*'s science writer. She previously worked in London as a journalist for *New Scientist* magazine, and her reporting has appeared in publications including *The Atlantic*, *VICE*, the *Sydney Morning Herald*, *The Age* and the *Saturday Paper*. Her work has been anthologised in *The Best Australian Science Writing 2020* and *2021*, and was shortlisted for the Bragg Prize in 2020. She holds a masters degree from the University of Cambridge.

CONTENTS

FOREWORD

Mary O'Kane

When historians look back on our time, they may call it the age of global information. First the internet, then social media, and now Artificial Intelligence products are flooding our world with data. Improved data sharing is fuelling scientific and engineering advances across a broad frontier, but not all information is knowledge. When it comes to drinking from the 'firehose' of information, not all the water is crystal clear – some can be murky, some can leave a bad taste.

This is a shame since there are so many jewels of exploration, discovery, creativity and invention to be enjoyed. The process of science has been embraced in Australia and across the world. School children love science, classes connect with the environment; and so many people develop a love of natural history. Citizen scientists engage with projects across society; interested people follow advances in technology. Informed and curious Australians enjoy debates on global affairs, politics and sport, but we also engage in rational discussions related to technologies in energy, water, food and health. And we all love the science of weather. During the Covid pandemic the quality of analysis and understanding shown in Australia was something we can be proud of.

But as information mounts and disciplines deepen in sophistication, it becomes more and more difficult for any one person to 'know it all' and be able to judge quality information from uninformed opinions, or worse – from deliberate misinformation. It also becomes harder to identify uplifting, thought-provoking insights from information dumps. In such times, busy people and those eager to learn new fields benefit from having the world's best guides who can shepherd them through the forest of data and encourage thought and wonder.

The annual anthology *The Best Australian Science Writing* is now in its thirteenth year. Each year authors submit pieces for consideration. And each year the quality and number of submissions continues to impress. Professional writers, scientists and science communicators send in essays that are carefully considered by the editor and a panel of experts. The final collection covers a snapshot of curated works that capture the issues of the day, as well as vignettes highlighting new and unexpected topics.

In a world of unrelenting pace, one can dip into this volume and read one or two pieces, or settle down on a vacation to read the whole anthology. One will see the quality of writing and the quality of work being done here in Australia, and stories about work being done right across the world. The book celebrates Australian science and our strengthening intellectual culture which has shrugged off the cultural cringe. It supports the communities who believe we can lead the world in calm, rational, practical thinking and the doing of science – and its translation through engineering, health and industry.

This year's addition of the *Best Australian Science Writing* is another step along the road of Australia's story, and a tribute to the people in our community who do and communicate great science.

Mary O'Kane

INTRODUCTION: A MATTER OF SCALE

Donna Lu

There is a line in the American journalist Elizabeth Kolbert's book, *The Sixth Extinction: An Unnatural History*, that I first read nearly a decade ago and which has stayed with me ever since: 'a hundred million years from now, all that we consider to be the great works of man – the sculptures and the libraries, the monuments and the museums, the cities and the factories – will be compressed into a layer of sediment not much thicker than a cigarette paper'.

It is an arresting image of the transience of human civilisation in the grand scale of geologic time – one that illustrates the distinctive power of science writing to provide perspectives that transcend the anthropocentric.

Science writing reveals to us new insights about ourselves, yes, but uniquely it also encompasses the vastness of the universe and all the diverse life forms within it. In the entries that make up *The Best Australian Science Writing 2023*, the scale of focus ranges from the microscopic to the astronomical.

It was humbling to read the incredible and incredibly varied output of Australian science writers from the past year, and whittling down the submissions was an extremely difficult task. There were many fine pieces which I unfortunately was not able to include.

The judgement of what constitutes the 'best' science writing is necessarily subjective. To borrow from the novelist Ian McEwan, writing in the *Guardian* on what might merit inclusion in a scientific literary tradition: 'Is accuracy, being on the right track, or some approximation of it, the most important criterion for selection? Or is style the final arbiter? ... We know what we like when we taste it.'

To me, the best science writing, as you will read in this anthology, is nuanced and intrepid, clarifies the complex and uncertain, and – perhaps most importantly – stems from an insatiable and infectious curiosity. It reveals fascinating insights into our surrounding environments and inspires awe at the wonders of the natural world – but, when necessary, also questions and highlights shortcomings of the scientific endeavour.

Science seeks to understand and address some of the biggest problems of our time: the existential threat of climate change, the worrying acceleration of biodiversity loss, the burden of human disease.

The impacts of climate change, perhaps the defining crisis of our era, reach vastly disparate ecosystems, from the tropics to Antarctica. **Jo Chandler**, in the piece that opens this anthology, reports on the urgent and ambitious Million Year Ice Core Project, which seeks to recover a million years of Earth's climate history by drilling deep into Antarctica's ice sheet. Bubbles of air trapped within the ice yield valuable information about environmental conditions – carbon dioxide concentrations, for example – in the deep past. Claude Lorius, the renowned French glaciologist who died this year aged 91, warrants a mention here. During a 1965 polar expedition, Lorius dropped some old ice into whiskey. Seeing bubbles of air sparkling in his glass, he realised they were samples of the atmosphere trapped in the ice, and conceived of the importance of ice drilling.

The icebreaker RSV *Nuyina* will play a critical role in monitoring the effects of climate change in Antarctica. **Jackson Ryan** sailed on its maiden voyage, and explains why the vessel is so critical to Australian Antarctic research. **Drew Rooke** highlights the tragic dieback of a species of cushion plant, *Azorella* *macquariensis*, on Macquarie Island, in a piece that emphasises the vulnerability of the subantarctic islands and their unique ecosystems to climate change.

The human impacts of the climate crisis are being felt perhaps no more painfully than by communities in the Pacific. **Helen Sullivan** travels to the Federated States of Micronesia, whose citizens face not only a combination of rising sea levels, drought and an acidifying ocean, but also a geopolitical tug of war between the United States and China. Further south, **Miki Perkins** covers a legal fight led by Torres Strait Islander elders against the Australian government, to protect their communities from going under – specifically by setting emissions reduction targets grounded in scientific evidence.

Humans are, of course, only one affected species among many. **Rebecca Giggs** writes poignantly about the erstwhile spectacle and the troubling disappearance of the migratory bogong moth, and her personal quest – as yet unsuccessful – to spot one. In Victoria's Central Highlands, towering forests of old growth mountain ash are threatened by fires and logging, and their survival, **Belinda Smith** and **Alan Weedon** explain, is critical to Melbourne's water supply. Australia has the dubious distinction as the world leader in mammal extinctions, and the mammologist **Euan Ritchie** explains what must be done to stem further biodiversity loss. One factor, he points out, is the persecution of our largest land-based predator, the dingo. **Zoe Kean**'s piece on the dingo fence, which stretches through south-east Australia, illustrates the ecological effects of dingoes' removal from the environment.

Despite cascading threats, species and ecosystems can surprise us with their resilience. **Angus Dalton** documents the remarkable transformation of the Macquarie Marshes, an internationally recognised wetland in Australia, after three years of rain. But its long-term future – or disgraceful decline – hinges upon political decisions that dictate water management practices.

In a joyous poem, **Felicity Plunkett** describes the resilience of the pink flannel flower, which 'labour[ed] in fire', eventually 'to come back – hailed by lyre, by whip – from / catastrophe'. O hope, indeed.

Sometimes, the survival of a species requires more drastic intervention. **Amalyah Hart** details an exciting but controversial approach researchers are taking to save southern corroboree frogs: arming them against the devastating chytrid fungus through gene editing. While the technology offers the potential for scientists to make precise changes within a single generation, there are also attendant ethical concerns regarding its use in conservation.

Other advances in genetics have had impacts on our legal system. **Nicky Phillips** explains how new evidence about gene variants has affected the case of Kathleen Folbigg, who was convicted of murder then pardoned 20 years later after science raised the possibility that at least two of her four children died of natural causes.

Though we have learned much about the human body, there is still much more to discover, as **Elizabeth Finkel** points out while reflecting on another decade of neuroscience research. Health conditions that primarily affect women are notoriously under-researched, and **Alice Klein** shines a timely spotlight on what scientists are learning about polycystic ovary syndrome, a common but long-neglected condition. Overseas, researchers are investigating the potential use of psychedelics in treating chronic pain, writes **Clare Watson**, who avoids the pitfall of 'hope and hype' by reporting on early-stage research with requisite caution. In a rich personal essay that draws upon both science and culture, **Heather Taylor-Johnson** conveys her experience of living with Ménière's disease, while after years of coronavirus lockdowns, **Paul Biegler** explores the potential health impacts of isolation.

Three years on from the first intrusion of Covid-19 into our lives, the virus has not gone away. A parliamentary inquiry into the health, economic and social impacts of long Covid and repeat infections released its recommendations this April. As **Bianca Nogrady** explains, the exact definition of long Covid remains a challenge, and managing the sometimes debilitating condition is difficult. Researchers are still exploring possible causes for long Covid, with a clarion call for us not to repeat past mistakes made with conditions such as myalgic encephalitis/chronic fatigue syndrome.

Looking to the future, scientists are investigating why bats are able to tolerate so many viruses that are deadly to people and other mammals, reports **Smriti Mallapaty**. It's hoped that the answers they find may help prevent the next pandemic.

Science, being a human endeavour, is necessarily fallible, and the field is improved by those who are unafraid to point to its flaws. **Jane McCredie** writes about the negative impacts of failing to include diverse populations in research, which limits the generalisability of findings and results in tangible disadvantages for minority groups. **Tabitha Carvan** sheds light on an academic's dogged quest to debunk inaccurate reports about the legendarily elusive night parrot, in a piece that raises questions about who we choose to listen to and the importance of truth in conservation.

A joy of editing the anthology was reading pieces that evoked sheer delight. **Lauren Fuge**'s was literally uplifting: she climbs 70 metres up a blue gum tree in Tasmania's Grove of Giants. The ascent into the canopy gives an awe-filled perspective on the interconnectedness of the planet and the systems it contains. Staying up in the trees, **Anne Casey**'s poem about tawny frogmouths evokes the spellbinding pleasure of birdwatching 'in stunned gratitude'. **Fiona McMillan-Webster**, meanwhile, details the intricate interplay between animals and plants – seeds, specifically – and their mutually beneficial co-evolution. Further afield, **Sara Webb**'s excitement is palpable as she explains what the launch of the James Webb Space Telescope means for astronomy.

It is, as **Meredi Ortega**'s poem 'First Light' succinctly puts it, a 'Starlight reveller, time / traveller, otherworld teller, eclipser of suns.'

With our focus still aimed skyward, **Jacinta Bowler** explores the significance of the SKA-Low project in the West Australian desert, which will form part of the largest radio telescope of its kind on Earth. The increasing number of satellites in low Earth orbit, however, poses a significant challenge to radio astronomy. **Alice Gorman**, a space archaeologist, discusses the growing problem of space junk, and the lack of an environmental management framework for space that would require major spacefaring nations and wealthy corporations to take greater responsibility for their activities. Having too many satellites in low Earth orbit also contributes to light pollution – an issue, **Karlie Noon** and **Krystal De Napoli** write, for astronomers, and First Nations astronomers in particular, because dark sky constellations play an important role in Indigenous knowledge systems.

And to go out with a bang, **Carl Smith** examines how scientists create antimatter, a mysterious and explosive substance which annihilates matter when the two come into contact.

I hope you derive as much pleasure from reading *The Best Australian Science Writing 2023* as I did from editing the anthology.

BURIED TREASURE

Jo Chandler

The photo that pops up with his periodic tweets is something of a non sequitur. Joel Pedro – 'Lead Project Scientist, Million Year Ice Core Project, Australian Antarctic Division' – sits at the wheel of a Massey Ferguson tractor that's seen better days, flannel shirtsleeves rolled to the elbow. Behind him there's a glimpse of pale acres of rye grass – fodder for cattle grazed on his late grandmother's property in Walpole, on the Western Australian south coast.

Click through to his profile and the farmyard shot is juxtaposed with one rather more in keeping with his polar credentials. Here he's perched on the wide tracks of a beast of an all-terrain vehicle in a dazzling icescape. Swaddled in a goose down jacket, he's smiling broadly behind a frosty beard and wraparound shades.

The two photographs riff on a remarkable journey – from tractor to polar tracks, third-generation farmer to glaciologist. It would seem fair to characterise this as an unlikely life trajectory. But google Pedro's hometown, and the tourist hyperbole would imply it was destiny. 'North Pole, South Pole, WALPOLE!' Still, what are the odds?

Conjuring a little more synchronicity, it's worth observing that the fortunes of Pedro's settler–farmer forebears were always at the whim of weather. He grew up through an era of declining winter rainfall across the region. And as is now understood – thanks to the work of the same glaciologists whose ranks he's joined – these conditions are entwined with mighty forces stretching across the Southern Ocean and deep into the ice. When circulating winds send moist, warm air down to East Antarctica – delivering higher snowfall to the coast near Casey Station – they tend to cycle back dry, cool air and create drought in south-west Australia. Ice core records indicate this strengthening pattern is likely not a natural event but a consequence of human influence on the climate.

And this is the business Pedro has found himself in, extracting relics of history from the ice, 'these really tightly connected components of the climate system – temperature, carbon dioxide, sea ice, ocean circulation – all so exquisitely and tightly linked together', Pedro explains. 'It's something that comes out of paleoclimate science in general. The closer you look, the more everything is linked together. And it only takes quite small changes to trigger cascading things.'

As an undergraduate at the University of Western Australia, he 'teetered on the brink of working on the salinity problem in WA'. But at forty-one, Pedro belongs to a generation weaned on warnings about rising temperatures. In his lifetime, levels of atmospheric carbon have skyrocketed. A fascination with atmospheric chemistry – 'not so much the white-lab-coat chemistry ... rather the more adventurous side of it' – prompted Pedro to apply on spec to the Australian Antarctic Division (AAD) for some postgrad work. 'They were looking for someone to work on reconstructing solar activity from ice cores using a cosmogenic isotope. And I kind of knew nothing about that, but thought it sounded pretty cool.'

Pedro drove his panel van across the Nullarbor, and twenty years and a few twists and turns later, he's leading Australia's full-throttle return to deep-field Antarctic science, heading its most ambitious and costly over-snow expedition in a generation. The objective is to set up a camp in the high interior of the East Antarctic Ice Sheet and drill a hole almost 3 kilometres deep. Over the next several summers, the crew will return to extract, catalogue and preserve 3-metre lengths of glacial ice laid down over a million years, pushing through the moment when something cranky, dramatic and mysterious happened. Entrenched rhythms in and out of ice ages blew out, catapulting the planet into a profoundly different state. Understanding just what happened way back then promises critical clues about conditions of life on the next Earth, the one human emissions are now conjuring into an ever-spiralling reality.

'What we're trying to do here is understand where the tipping points are in the climate system,' says Pedro. What was going on in the atmosphere in the past, in particular with greenhouse gas levels? What was the solar story? If we have this information, he says, 'then we have a firm handle on these guardrails of where the climate system is stable and where it tips. And obviously what we're looking at for the future is how far we can push the climate system before it tips into another state.'

Australia's million-year ice core project is unequivocally a mission of discovery, and an urgent one. But the optics are also unambiguously strategic as Australia muscles up its Antarctic credentials and influence. The revival of the AAD's long-mothballed deep traverse capability – the equipment, logistics and skills necessary to operate long-haul expeditions on the ice – is just part of a multi-billion-dollar polar science program which, when it was signed off by Canberra in 2016, emphasised its service to the national interest and international relations.

In the annals of Antarctic law and lore, there's a good deal of reflection on the motives of science, with the upshot that only the most naive or cynical could fail to grasp their entanglement with politics and strategic posturing. The remarkably resilient Antarctic Treaty – signed in 1959 and brought into force two years later – preserves Antarctica for science and peace, putting all territorial claims on ice and fostering authentically warm collaborations between scientists whose nations can include the frostiest of foes. But it can't cleanse the continent of national agendas. And as an American expert noted even as it came into force, 'whatever advances science furthers strategic techniques: a station useful for gaining knowledge of our environment is ultimately strategically important by its very nature'.

China is right now also busy drilling for the prize of oldest ice, as are Europe and Japan. Russia is in the game, and South Korea has plans. Australia's program, 15 years in the making, has long been at the forefront, but Pedro's team has been delayed for two precious summer seasons by the global pandemic and poor luck with the weather. Any of these programs may stall or pull up short. But the hope is that at least a couple of them will retrieve the oldest ice in the next few years – more than one being ideal, to validate and replicate findings.

A million years to go, and no time to lose.

The wealth hidden within the polar ice has long been suspected, if only lately understood and exploited. In 1894, returning from a whaling expedition, Scottish artist and explorer WG Burn Murdoch wrote of 'the mysteries of the Antarctic, with all its white-bound secrets still unread, as if we had stood before ancient volumes that told of the past and the beginning of all things'.

Over the ages, snow falls on the polar ice, accumulating year on year, deeper layers compressing under the weight of fresh flurries. When the compacted snow turns to ice, bubbles of air are trapped. Ice cores mine this treasure, bringing to the surface vials of atmospheric history that scientists carefully haul back to the laboratory to prise open, like breaking the seal on an ancient tomb.

From these samples, scientists can discern past air temperatures, measure precipitation and track concentrations of carbon dioxide, methane and other gases. They can open the pages of deep time through traces of aerosols and micro-particles. Vestiges of ash archive volcanic eruptions, salts recall conditions in the surrounding seas, sulphur signatures etch the extent of sea ice and the life forms clinging to it, sprinklings of mineral dust testify to wind circulation. This isn't proxy data – that is, the indirect record of deep time constructed out of tree rings and shells and the like, which is hugely valuable but often maligned by the anti-science brigade. Nor is it modelling, which is similarly undermined. What's in these bubbles is the real deal.

The Soviets began drilling for old ice in Antarctica in the 1970s, burrowing determinedly for over 30 years. In January 1998, they pulled up a core 420 000 years old from beneath the remote Russian research station of Vostok, near the centre of the East Antarctic Ice Sheet. By then the Russians were collaborating with American and French scientists, reconstructing atmospheric and climate history from core samples stretching back through four glacial–interglacial cycles.

When they published their findings in a watershed paper in the journal *Nature* in 1999, it transformed understanding of our planet and human influence on it. Interviewing the eminent climate scientist and activist Will Steffen some years back, I asked if he recalled the galvanising moment in his scientific journey. This was it. 'For the first time we saw this beautiful rhythmic pattern, how the Earth as a whole operated. You saw temperature, you saw gases, you saw dust all dancing to the same tune, all triggered by the Earth's orbit around the sun … but still a mystery, it couldn't explain the magnitude of those swings.

'To me the Vostok core was the most beautiful piece of evidence of the Earth as a single system. We spent a week trying to understand the data, but the only way you could make sense of it was to recognise the strong role of biology. Before then, the earth was seen as a big geophysical system … with life at its whim. But life is actually an important controller, a strong influencer of what went on.'

Glaciologists kept drilling deeper and further back in time. The European Project for Ice Coring in Antarctica (EPICA) got to work near the French–Italian inland station of Concordia back in 1996–97. Their first hole at Dome C had to be abandoned in 1998 when the drill got stuck. So they moved the rig 10 metres north and started again, returning each summer to push deeper until December 2005, when the drill began approaching bedrock.

They stopped at a depth of 3260 metres, when geothermal heat rising up from the Earth started melting the deepest reaches of the hole.

'They got to 800 000 years,' says glaciologist Tas van Ommen, the AAD's climate program leader and co-chair of the International Partnerships in Ice Core Sciences (IPICS). But 'darn it, they didn't solve the problem'. Van Ommen is not one for hyperbole.

'The problem' is the Mid-Pleistocene Transition (MPT). We've known since late last century, from marine sediment cores, that from 3 million years ago until about a million years ago, the Earth swung in and out of ice ages like clockwork, each cycle lasting around 41 000 years. And this fitted sweetly with the century-old hypothesis of Serbian scientist Milutin Milankovitch, who had calculated that ice ages would occur every 41 000 years based on the tilt of the Earth's axis of rotation as it travels around the Sun. Then the cycle changed, and by 800 000 years ago it had blown out to a new pattern of 100 000 years. This fits with another cycle described by Milankovitch, tracking changes in the shape of the Earth's orbit over time from nearly circular to slightly elliptical. The planet was now dancing to a different, slower tune, as if someone had dialled the turntable down from 45 to 33 rpm. 'It's really worrying as a scientist when you realise you could have a perfectly good explanation for either [cycle], but you can't explain why it would change,' says van Ommen. 'That is the massive question behind the million-year ice core.'

There are several theories. One points the finger at surface changes, theorising that the advance and retreat of ice at the quicker, pre-MPT rhythm scoured back the regolith – the scatter of loose rocks and dust sitting above bedrock – which oiled the flow and spread of the ice sheets. Without it, the ice stuck and grew higher and less vulnerable to melting. Another argues that the Northern and Southern Hemispheres were cycling in and out of ice ages at different times and then somehow fell in sync. Then there's the idea that carbon dioxide was steadily declining through this period, cooling the climate and bulking up the ice sheets. With a million-year-plus ice core, the mystery could be solved.

The EPICA drill pulled up just short.

When he explains the histories locked inside ice cores to politicians, policy-makers or journalists, van Ommen works methodically through a set of PowerPoint slides. He starts with Confucius – 'Study the past, if you would divine the future' – rendering the graph that comes at the culmination of the show all the more magnificent and terrifying.

This graph is plotted against the EPICA core record, drawing out two threads of data retrieved from its ice. A black line tracks the atmospheric carbon dioxide (parts per million) and, above it, a red line traces Antarctic temperatures. The lines pulse together in peaks and troughs, like the vital signs of a patient hooked up to a monitor. They are almost entirely in sync, temperatures rising and falling to the same rhythm as carbon dioxide. 'When you see them together, you say "Wow!",' says van Ommen. 'You can argue with the sceptics about which leads and which lags – it just depends on who is doing the forcing. But they are connected, sort of like a rubber band, one will pull the other.'

In the final, squeezed fragment of the EPICA graph the carbon dioxide line shoots violently upward into uncharted territory. Over the 800 000-year record, atmospheric carbon dioxide has never peaked over 300 ppm. For all human history, it sat around 275 ppm until about 200 years ago, when we began to burn coal to fuel the Industrial Age. In 1950, it punched through the 300 ppm historic ceiling. In mid-May, as the forests of the Northern Hemisphere dropped their leaves, the planet exhaled atmospheric carbon dioxide at a new daily record of 420 ppm. This is where a patient's machine would sound a piercing alarm and emergency teams would materialise bedside stat!

'There's something really fundamental we don't know about the planetary system,' says van Ommen. This 'something' being the on–off switch on ice ages. 'As far as I'm concerned, knowing fundamental things about our spaceship – Earth – is almost a societal issue in itself ... But in terms of practical "what do we need to know?", it's about informing us of the long-term risks of what we're doing and where the tipping points or commitments might be.' And should humanity get its (our) collective shit together, this will inform whether there is scope to overshoot for a century or two while we get into negative emissions territory. 'It really is a burning question in planetary climate science.'

Back in 1957, after investigating the question of whether the oceans had enough appetite for carbon dioxide to soak up rising fossil fuel emissions, American oceanographer Roger Revelle concluded that they did not. As a consequence, he wrote, greenhouse warming 'may become significant during future decades'. This meant that 'human beings are now carrying out a large-scale geophysical experiment of a kind that could not have happened in the past nor be reproduced in the future'.

Today, despite all such warnings, and with humanity still disinclined to pull the levers on what's playing out in our real world laboratory, van Ommen enlists the same analogy. 'And since we're doing a really big experiment, it's kind of good to know the answer.'

In October 1994, glaciologist Mike Craven and five other men – three diesel mechanics, a surveyor and an electronics engineer – chugged out of Larsemann Hills, a field site 110 kilometres south-west of Davis Station, aboard three tractors towing vans, generators, food and fuel. They set course for the world's largest glacier, the Lambert. Over the next 120 days they stopped at 72 locations spread 30 kilometres apart over a 2250-kilometre survey route ending at Mawson Station. Top speed on travelling days was 5 kilometres an hour, working in temperatures down to minus 40 degrees Celsius, pulling up to collect measurements and ice cores, tend and refuel the tractors, siphon their waste into the empty fuel drums, wash and eat and sleep, and do it all again.

Theirs was the fifth – and final – annual AAD traverse to the Lambert Glacier basin, tracking 'the movement of the ice; not just the amount but in what direction, the velocity', Craven explains. Because inland ice flows so slowly – maybe 10 to 20 metres a year – readings had to be precise. Every 24 hours they clocked GPS positions continuously for 12 hours at the survey sites, which was the fastest the technology of the day could manage with the fine accuracy required. Every 2 kilometres, following the trail of previous teams, they would come to a cane driven into the ice and measure snowfall over the previous year. Periodically they would find a cane farm – 100 canes planted in rows 10 by 10, set 20 metres apart – back on the first traverse to get a fix on local accumulation. 'And you're measuring with ice radar every day, so you get the thickness of the ice and the velocity, which means you get a flux of the ice across a given section of track.'

Every week or so, they would drill out a 25- to 30-metre ice core. Craven would split it down the middle, saving half to haul back to Hobart for laboratory tests and archiving and getting to work on the other half himself. Using a high-voltage current, he could distinguish layers of summer snow from traces of sea salt. Or he could put the core on a light box and read the seasons in bands: summer ones translucent from the surface melt; winters opaque with bubbles left by snow blowing over the surface. From density measurements he could calculate how long it had taken the snow to turn to glacial ice. In coastal areas, the angle of the sun on the slope means more melt, and snow can turn to glacial ice – locking in air – within a hundred years. Inland the process might take 5000 years. Glaciologists need to figure out the timing at every site to accurately ascertain the age of air trapped in an ice core.

The journey ended 'about February 17 (1995) – I know because that's my wedding anniversary', Craven recalls. We're at his home in Hobart, where he and wife Chris are shortly expecting some old Antarctic hands for a barbecue. He's dug out his diaries for our interview but has barely glanced at them in nearly two hours with my recorder running. I recall another Lambeth veteran once telling me 'there was a Zen-ness about it ... Every minute there were changes in the snow surface and you could detect them. You could feel the degree of change in the slope. You become so desensitised to all other distractions you develop another acute observational skill ... Antarctica is like a drug. It's like a love affair.' Arriving at Mawson was like 'arriving in paradise', Craven says. 'Mawson had these wonderful mountains coming up through the ice. I basically walked the last one hundred kilometres ... on cross-country skis. It was just magnificent.'

It was also the end of an era. Australia's Lambert Glacier traverses are recognised among the most extensive over-snow science programs undertaken on the continent between 1950 and 2000. But they were a last hurrah, with the tempo of the legendary Australian National Antarctic Research Expeditions (ANARE) activities petering out since the early '80s as attention switched to building programs, budgets tightened, and aircraft and satellites were increasingly enlisted to collect data. After the 1995 run, 'those traverse vans and vehicles got distributed around the stations, pushing snow out of the way, doing odd jobs, not really doing any traversing', Craven recalls.

There's a truism in Antarctic circles which observes that 'science is currency'. The line is often credited to a 1989 paper by Tasmanian political scientists Richard Herr and Robert Hall, which documented Australia's waning enthusiasm for Antarctic science and expeditions, and its growing emphasis on occupation of the ice, building living quarters, amenities and services. It cautioned that in the event of any review or ructions in the treaty, these priorities might not prove particularly useful.

Antarctic spending was still deep in the doldrums 20 years later when I visited as a reporter, accompanying AAD scientists to Casey in 2007 and 2009–10. Their distress back then about eroding budgets and the fallout of their research on urgent climate questions – the stability of glaciers and implications for sea levels, ocean warming and acidification, the security of krill and fisheries, the vulnerability of penguin colonies – was at fever pitch. With so many berths occupied by tradies and support crew, competition to get to the ice to do fieldwork was intense. Contemplating the ravenous fuel, food and logistics costs of maintaining station operations, one senior scientist grumped that he would give up the comfort of a warm station for a tent on the ice in a heartbeat if that freed dollars for actual science.

'There was this period – from the '80s through to really 2010 and onwards – where Australian governments lost interest ... the investment tailed off,' says AAD chief Kim Ellis. That, and an increasing emphasis on safety protocols, was the undoing of the era of 'men with beards doing adventurous things'. In retrospect, Ellis witnessed the turning of the tide during his first trip south in 1979–80, when as a young soldier fresh out of Duntroon he volunteered for an amphibious detachment running resupply operations as the building phase gained traction. Since taking the helm at AAD in 2019, he's been in the box seat to see priorities dramatically swing again.

On 16 October 2021, the single biggest investment in the history of Australia's Antarctic program materialised out of the mist and sailed up the Derwent River, pausing to show off with a couple of 360-degree twirls and a blast of its horn. The icebreaker RSV *Nuyina* was described as 'Disneyland for scientists', bristling with sensors, cameras and hydrophones, tanks for scooping up seawater and krill, a 'moon pool' dropping from the science deck down through the hull, allowing launch and retrieval of equipment in any conditions. The vessel cost $528 million to design and build, which sits within a $1.9 billion package to maintain and run it over the next 30 years.

Meanwhile, the federal government has in recent years committed $450 million to upgrading Casey, Mawson and Davis stations, $50 million to build a new one on Macquarie Island and $77 million to scope out a year-round aerodrome near Davis –

a proposal scrapped last year on environmental grounds. Then there's the $52.5 million earmarked to revive the over-ice traverse capabilities that will, as their opening act, be used by Joel Pedro and team to get to Dome C and start drilling. In February 2022, the then prime minister Scott Morrison announced another $804 million for strategic and scientific programs over the next decade, including drones, helicopters and other vehicles allowing exploration of inaccessible areas and an 'Antarctic eye' remote monitoring program.

Plainly Australia has developed a keen appreciation of the value of science as currency in matters Antarctic. Kim Ellis pegs the turning point at around 2011, with growing global interest in Antarctica and its frontline exposure to climate warming, not least in regard to the sustainability and management of Southern Ocean fisheries. The case for the investments flowing now were laid out in a 2014 report by Tony Press, a former AAD director, who argued that Australia's standing was eroding as a consequence of its underinvestment just as new players were emerging on the ice. 'The leadership that Australia has naturally assumed by its proximity, history and experience risks decline.' The preservation of the ice and ocean under the terms of the Antarctic Treaty System (ATS) was powerfully in Australia's strategic interests, Press argued. Alongside this was the drumbeat of geopolitics – notably China's rise and escalating Antarctic program, and questions around Russia, both big polar players.

In 2016, the Turnbull government signed off on investments from the *Nuyina* down to the million-year core project in the Australian Antarctic Strategy and 20 Year Action Plan, building on many of the recommendations of the Press report. It was, says Ellis, a simple but transformational document – 'not a lot of nations have that clarity in what they want to do in Antarctica'. In the foreword, Turnbull celebrates Australian expeditioners back to Sir Douglas Mawson – 'a legacy of heroism, scientific endeavour and environmental stewardship'. The government was 'delivering a new era of Australian Antarctic endeavour. It is timely to reaffirm our national Antarctic interests, and to set out a plan to protect and promote them.'

One of my craziest, fondest Antarctic memories is a hike from a field camp at Bunger Hills to a ghostly station built by the Soviets in 1955. Deep in the chill of the Cold War, they had launched the first Soviet Antarctic Expedition. Whatever Moscow's motives, the scientists of Oasis Station were intrigued by its ice-free lunar landscape and lake. They measured ozone and solar radiation. They explored glaciology, seismology, geomagnetism, biology, hydrology, auroral physics and gravimetry before handing the building over to Polish scientists. Since 1979, the base had been occupied only sporadically. At the dawn of 2010, it was a mess of broken glass and vintage technology. There were cigarette butts, empty vodka bottles, a gramophone and a stack of 78s with 'CCCP' stamped on red labels. Behind a hammer-and-sickle flag was a musty bunk room, and on the ceiling a message: 'Dale Andersen. Joint US/ Russia Bunger Hills Expedition 91–92 – Onwards to Mars!'

When I tracked him down, Andersen was working at the University of California's SETI Institute. An astrobiologist, his interest at Oasis was in the lake, which exists at the very boundaries of conditions for life – a proxy for Mars long ago, when the red planet's climate may have nurtured basic life. The trip was a great success scientifically. And the friendships and collaborations endured 20 years on. He sent me pictures of the US and USSR flags flying over the decrepit station and of him and Sacha, the girl he fell in love with on the icebreaker, who remained his Russian bride. Here in this remote place, scientists from a divided world had worked together. Antarctica is full of such stories – about Russians and Americans in the depths of the Cold War, British and Argentinians through the Falklands conflict, Chinese, French, Japanese, Koreans – you name it.

The episode encapsulates the disparate threads of the Antarctic narrative that are so captivating and entwined but that resist the constraints of the usual polar tropes: derring-do, geopolitics, nature, science. The individual quest for knowledge, recognition and adventure. The exposure of intimate character to the elements, the layers of veneer stripped back to raw. The chest-thumping of posturing nations, the camaraderie of their emissaries. The outposts of our occupation of this place, which defied human intrusion for so long, are still so tenuous. Rusting fuel drums piled on white ice. The juxtaposition of mess and purity, vulnerability and magnificence, transience and endurance. This is Earth before and beyond us.

So at the risk of joining the naive or cynical, I'm curious about the bottom line of Australia's exploding Antarctic investment – is it about politics and geostrategic one-upmanship, or about science and discovery? Maybe I'm just a sucker for the notion that the pure polar conditions might nurture our better angels. Tas van Ommen has no qualms about recognising the *realpolitik* of Antarctica – that it can be both, and for all the right reasons: 'The hope is that you capitalise on the other imperatives for being in Antarctica, and the fact that the treaty says that science is at the centre of why we're there ... we are now stating very publicly that we are there for science.'

'So, Jo, the correct answer for me to give, of course, as director of the AAD, is that that is a decision for government,' says Ellis. 'Probably a more pragmatic answer is that of course governments are going to respond to what they think are our challenges, what impacts on us. We in the Antarctic Division have a responsibility to deliver against government priorities. And if government priority is about presence and influence in Antarctica, that's part of my job as well.'

What it's not, he adds, is about any sense that 'we've got to securitise Antarctica. I think that's a huge mistake. And there are certainly commentators who are arguing that ... What the current paradigm has been built on is this idea of shared research, peaceful intent, compliance with the protocols of the Antarctic Treaty. And as soon as we move away from that we take a real risk.'

Ellis talks about a trifecta of currencies of influence in Antarctica – science, presence and excellence in operations. All three are now more funded and better demonstrated, 'and they allow us to deal directly with other nations to ensure that we have some say in what happens in Antarctica ... I'm very cautious about anyone who talks about, you know, toughing it out with other nations. That's just unhelpful and it is not in the spirit of the work we do.'

At which point it's well beyond time to acknowledge the panda in the room. And the bear as well.

On 11 January 2010, days after my trip to Oasis and winding up a wild week of whale surveys from Bunger Hills, my Antarctic adventure came to a frenzied finale. A worker at China's Zhongshan Station had been hit by an earthmover and horribly injured. He was flown to the Russian station Progress II, where Australian doctors from Davis and Casey dashed to assist. A medical team from Hobart was scrambled onto the weekly Antarctic airbus shuttle – my ride out. The patient was stretchered aboard and into their care behind a curtain. Three nations engaged in a life-and-death drama to save one precious life. I never heard what became of him. Again, Antarctica is full of such stories.

In December 2020, the shoe was on the other foot, with Chinese and US teams carving ice runways to medivac a sick Australian out of Davis, even as trade and pandemic tensions raged between Washington and Beijing, Canberra and Beijing. For the most pragmatic of reasons, cooperation on logistics and rescues has been ever thus, says Tony Press. We chat in Hobart in December 2021 as Australian–French diplomacy is spectacularly tanking, yet submarine chat is a mere *amuse-bouche* when French and

Australian polar heavyweights meet for a festive dinner. 'Keeping those levels of logistic and scientific collaboration when there are other tensions, regardless of whether they're Antarctic tensions, or trade tensions, or human rights tensions … it's one of the best forms of diplomacy,' says Press. 'It gets you the back-door channel to resolve issues before people start throwing bombs at each other. Which is where the Antarctic Treaty came from – to stop people throwing bombs at each other, fighting with each other over territory.'

Security analyst Claire Young argued in a 2021 Lowy Institute paper that China was pushing the ATS boundaries and 'wants to benefit economically, and potentially militarily, from Antarctica'. Her take on the treaty is rather more cynical than that of Press – she views it as a 'Cold War relic' that, stripped of its idealistic, scientific foundation story around the 1957–58 International Geophysical Year, was a pragmatic exercise by superpowers to cauterise a potential new frontier of hostilities. Nonetheless, it's a better deal than anything that might be wrangled today, she says, so it deserves support.

There's compounding action and interest down south. China is linking its Antarctic ambitions to its rising global profile, says Press. It's investing heavily in Antarctic fisheries and tourism, and there are dark mutterings around its potential interest in minerals – mining is banned under the treaty. It's active and sometimes disruptive in ATS forums. 'Then there's the re-emergence of Russia as an unpredictable player. And that brings new elements into the Antarctic diplomacy framework.' And this was before the Ukraine invasion earlier this year detonated a whole new dimension of uncertainty around Russia's activities at both poles. Meanwhile, China and Russia are opposing three new marine protected areas that would expand protections to about 20 per cent of the Southern Ocean.

All this feeds an increasingly militarised analysis of Antarctic behaviour in political and academic spheres, alongside grim soundings on the strength of the Antarctic Treaty to withstand rising pressures and on the prospects of the consensus process at its heart breaking down. Like Kim Ellis, Press and long-time Antarctic scholar Marcus Haward push back on this narrative on both its merits and its dangers. 'I think they are writing off the Antarctic Treaty too soon,' says Haward, who is co-lead of funded research investigating historical geopolitical stresses and their impacts on the ATS. 'We're creating perhaps a self-fulfilling prophecy by sort of looking at all the negatives. When we look back over the past fifty, sixty years, we see the consensus-building and collaboration actually often comes in these times of stress – always does, in fact.' He also says that through times of internal debate and external stress on the ATS, 'Australia's been there, and been a leader'.

Press thinks the treaty is solid but warns against complacency. The ATS relies on consensus, and China and Russia's current objections to expanded marine protection zones illustrate the hitch with that. 'The way things fall apart is corrosion. And one of the forces of corrosion is a set of practices develop and become common practice. One of those might be countries feel comfortable in just saying "no", not trying to reach a consensus on something. If that becomes common practice, the whole basis of the Antarctic Treaty System is corrupted.'

Haward argues that, if anything, the treaty is showing its viability as new countries join and show an interest. 'Why are they doing it? It's the science. It provides them with an opportunity ... [they] can see collaboration through science as a way of engaging internationally.' Part of his current research is looking at how the system coheres and how national interests are projected in Antarctica. 'They can be coincident, but they can sometimes diverge ... some of [China's] assertions perhaps are pushing some of these norms in a way that could fracture the system. I think that activities like the million-year ice core are symbols, and they are flag-raising exercises, of course, but it's saying, "Antarctica matters to us". The critics will say this is just protecting a claim – yes, it is.

It's saying that we take our territorial claims seriously. But we also take the Antarctic Treaty seriously, and that is about science. There are enormous spin-offs from the million-year ice core.'

When I first interviewed Joel Pedro it was mid-December 2021 and he was locked up in a Hobart hotel for sixteen days of quarantine – the protocol for all AAD expeditioners in the time of Covid. From here he made the heart-in-mouth call on the final coordinates for the drill site at Dome C: 122.5209° E, 75.34132° S. It was the culmination of years of scoping and surveys, the end of a process Tas van Ommen was beginning back when Pedro was his PhD student. By now it was narrowed down to a question of metres, informed by intimate scrutiny of the underlying bedrock to identify the 'sweet spot' most likely to yield the prize of oldest ice. 'And I was sitting here in this room and shouting [through the email] were the lead ops guys.' The French crew at Concordia Station was helping get some of the Australian equipment on site. 'They needed our coordinates to drop off our bore-hole casings. I'm writing "Guys, do we need to buy more time?"' But time – this season – was up.

On 17 December 2021, Pedro tweeted with excitement: 'We're coming for you old ice … just boarded flight Antarctica's Wilkins Aerodrome to join the first field season of @AusAntarctic @MillionYearIce project … busting to get on the ice and get out to Little Dome C with the team.' The core team waited at Casey for seven weeks for all the stars to align, always a high-stakes business given the variables and extremities of this place. In the end a combination of poor weather, limited operational windows and Covid scares at other bases thwarted any prospect of starting the drill for another year.

While they were waiting, the team did manage a run out to Law Dome, a small icecap rising almost 1400 metres to the east of Australia's Casey Station, to test their equipment and do a little coring. Renowned for the fine-grained clarity of the 750-year record it preserves, Law Dome has been invaluable in informing understanding of Australia's climate through countless scientific papers, including the one linking East Antarctic snowfall with the drought Pedro grew up with in south-west Australia.

Australia's relaunched traverse team looks rather different to those of the ANARE era. Two of the its members are women, including field leader Sharon Labudda. Originally from Kingaroy in Queensland, a secretary turned diesel mechanic, she started out in coal mines and is now a 20-year Antarctic veteran. She's spent 18 summers on ice and is currently in the middle of her third winter. When Pedro and the other team members flew home, Labudda stayed at Casey to start priming and equipping the newly delivered traverse vehicles to be ready to roll next summer.

Labudda has some serious traverse miles under her belt. She has ploughed her way from the coast into the Aurora Basin (once) and Dome C (twice) with the French program. We're talking 12 hours a day over 25 days each way, driving a tractor or snow groomer, swapping the wheel every hour or two 'because it's quite draining, especially in bad weather when you can't see much'. Temperatures were mostly –20 to –30 degrees, 'but we got down to minus 50 for a few days at the end.' She was sharing tiny bunk vans with men whose only common language was in the quirks and mechanics of the vehicles they drove. She did try to teach herself French, but it never quite stuck. 'I didn't realise how much I hadn't spoken until I got back after five months, and when I did talk I was talking really slowly and people were saying, "Are you okay?"'

As always, tracking back to the expeditions of legend – by Ross and Wilkes and Dumont d'Urville, Scott and Amundsen, Shackleton and Mawson, Byrd, Nansen et al. – the scientific mission in Antarctica is enmeshed with, and defined by, intimate exploration of the human spirit in extremis, testing the limits of bravery, capacity, morality. As a field leader, Labudda has had to become as proficient at reading, directing and repairing people as equipment. What's she learnt, I wonder, from observing humans working and living in such challenging conditions? She pauses a moment to consider, as is her practice in confronting pretty much any situation, as the job requires. 'That you can never judge a book by its cover.'

I put it to Pedro that it's more than the degrees of cold and the isolation that make the search for oldest ice daunting. It's the idea that you are on the absolute frontline at this seismic moment of understanding around the Earth system, and our influence and our future. It must be like blasting into space on the nose of a rocket. How do you manage that knowledge, that burden?

'It can feel like walking off a cliff,' says Pedro. 'If you engage with it too deeply it can be consuming, and I think even counterproductive.' Time with his kids, aged three and five, has helped, and his partner is also a scientist, an oceanographer. 'It's a bit frightening when you think that in the space of my lifetime, carbon dioxide levels in the atmosphere have changed as much as they do between a glacial and interglacial' – between Ice Age cycles of ice advancing and retreating. We're talking about a temperature change of 6 degrees that 'just transforms the planet – changes sea levels by 120 metres.

'And then you see that the pace of change is now 100 to 1000 times faster than those natural changes. It's just so bloody obvious – we can measure this in the carbon dioxide data, when you look at the isotopes. It's clearly fossil fuel burning. And then you see the state of some of the public and political debate, and that's a big worry. And it has been that way for too long now.'

Returning to Hobart without having made it to Dome C last summer, Pedro was disappointed but pragmatic. 'It kind of comes with the territory down there. It's a five-year project – we just don't want too many years like this.' The European oldest ice project – Beyond EPICA – has now completed its pilot drilling at Dome C and has the advantage of a nearby station, Concordia, and a well-established traverse capability. They'd be a strong bet to be the first to retrieve oldest ice, earning some pretty substantial scientific bragging rights.

But that won't be the end of the story. At least one more core will be needed to verify the first, to be sure that what's discovered is not some quirk or anomaly. 'This underlines why it's certainly not in our interest to try to make it a race,' says Pedro. 'The thing I'm going to be pushing is that we need to do things at our pace, do things well and sensibly, and to distinguish ourselves on the science side.'

When the million-year core team made their part-training, part-consolation visit to Law Dome last summer, Pedro writes me in an email, he 'twisted Sharon's arm to turn our diesel generator off for a while, so we could listen to the silence. We all sat outside then for some time and got kind of recharged by the vastness and the silence. Stunningly, while we were sitting there, a Wilson's storm petrel straying a good way from the coast flew through and pooed on me. I kid you not. We all had a good laugh about that, and decided it was quite lucky and somewhat symbolic of the season.'

Postscript

Kim Ellis is now the former chief of the Australian Antarctic Division. Will Steffen died on 29 January 2023, aged 75.

The ten members of the AAD million year ice core team set out from Casey Station on 23 December 2022, arriving at Little Dome C on 17 January 2023. Having established a safe route, proven the traverse capability and tested equipment, they returned to Casey on 11 February with shallow test cores. The return trip covered 2300 kilometres in temperatures down to −44 degrees Celsius. Next summer they plan to return to begin drilling for oldest ice.

✱ *A subantarctic sentinel*, p. **85**
Onboard the space station at the end of the world, p. **283**

THIS MAGNIFICENT WETLAND WAS BARREN AND BONE-DRY. THREE YEARS OF RAIN BROUGHT IT BACK TO LIFE

Angus Dalton

Biologist Sophie Hewitt is gunning an army-green, eight-wheeled amphibious vehicle not much larger than a quad bike across boot-sucking mud and pools of chest-deep water.

We're closing in on a distant burble that sounds like highway traffic. But it can't be; we're in the Macquarie Marshes, north-west of Dubbo, surrounded by 200 000 hectares of flooded wetland.

Dodging trunks of inundated coolibah trees and giant tangles of vine-like lignum, we get close enough to send up a drone. The video feed reveals the sound's source: a vast colony of honking, flapping, foraging straw-necked ibis. In each nest, on islands of lignum, are the specks of white eggs.

These thousands-strong waterbird breeding events are part of the reason the Macquarie Marshes are an internationally recognised Ramsar-listed wetland. (Ramsar is an international convention for protection of wetlands.)

'There's undoubtedly a full-on breeding season happening,' Hewitt's PhD supervisor, ecologist Professor Richard Kingsford, says. 'Because there's so much water in the system, there'll be a lot of colonies and they'll go for a long time, which is pretty unusual. It's just a frenzy, basically.'

This could be the most spectacular year in the marshes in three decades. The question is how far three back-to-back La Niña events and this year's record rain will go in restoring a wetland constricted by dams, scorched by record drought and threatened by climate change.

Hewitt, Kingsford, biologists from the Australian Museum and a gaggle of ecology students from UNSW have slipped into the marshes for a 36-hour window between flood pulses to help find out.

Between 2017 and 2019, the marshes were sapped by the driest three years on record. River red gums died and dried to bone-white. In an aerial survey, Kingsford spotted a single pair of black ducks sitting in dusty, crackle-dry reed beds.

Now, Kingsford estimates that at least 500 000 mega-litres, equivalent to the water held in Sydney Harbour, has gushed across the system. Between 200 000 and 400 000 megalitres usually kick off a major breeding event. It feels like every inch of the wetlands is soaked and thrumming with life.

Later, we wade into a forest of river red gums. Every fork is stuffed with a loose assemblage of twigs frizzed with eucalyptus leaves: the nests of rare, nocturnal night herons. Kingsford, who has studied the marshes for 37 years, hasn't seen them in the marshes for six years.

A whistling kite is raiding the nests for unguarded chicks. Behind us, a rakali (a native water rat – picture an aquatic ringtail possum) wends through the slow-moving water. Deeper into the trees are a few thousand breeding egrets.

Back at the road, students are recording sightings of pacific herons with 1.5-metre wingspans, emus ushering clusters of striped chicks across the floodplain, royal spoonbills crowned in their breeding plumage and the fluorescent blue-green flash of mallee ringneck parrots. The marshes also support endangered species including the Australasian bittern and painted snipe.

At night, we find broad-palmed rocket frogs, named for their turbocharged leap, and salmon-striped frogs, which sport pink eyeshadow. Frog biologist Dr Jodi Rowley points out cattle prints in the mud where spotted marsh frogs have laid their frothy eggs.

She's hoping to find a bucket-list species: the elusive underground holy cross frog. It looks as if Fabergé bedazzled a lemon, and only emerges during downpours.

'For a frog biologist, it has been absolutely heaven,' Rowley says. 'I have never seen so many frogs in my life.'

Aim your torch at the sky, and the reason for this explosion of life reveals itself: the static of moths, midges and mosquitoes is thick enough to blot out the Milky Way. Plunge a net into a roadside puddle and it will come out wriggling with tadpoles, mosquito larvae, water beetles and finger-length predatory dragonfly nymphs armed with spring-loaded retractable jaws. Every litre of soil can squirm with up to 10 000 microcrustaceans.

These invertebrates are the 'engine room' of biodiversity in the marshes, says Kingsford.

'I find it almost intoxicating to be able to go out into that environment and see how much is happening,' he says. 'But the scientist in me wants to look at the data and see how it compares to where this place has been in the past.'

The marshes once spanned a million hectares before the construction of the Burrendong Dam on the Macquarie River, which feeds the marshes, in 1967. The marshes constricted to one-fifth of their former size as water was diverted to irrigators (cotton is the primary crop in the marshes). The wetlands once flooded every two years. After the dam, it's more like every five years.

The number of waterbird nests in the marshes has plunged by 11 000 every 11 years since the dam's construction. In 2016, a big flood year, Kingsford counted 30 000 nests – a disappointing number. But the NSW environment and heritage group estimates there were 90 000 nests by February 2022. They estimate that number could have climbed to 150 000.

The number of nests counted over summer this year will give crucial insight into just how resilient this boom-and-bust ecosystem really is.

'The marshes used to come back like a super ball, and bounce back to the same height every time you had a big flood event,' says Kingsford. 'But now I think of these bounces as being like a tennis ball – the bounces aren't as high as they used to be.

'That will be the really interesting thing to me. How big a bounce have we got out of this one?'

Climate change will sap the marshes of more water in coming years but, for the time being, water extraction is the main reason for the marshes' downward trend. In 2009, the government notified Ramsar that the Macquarie Marshes were degrading from a semi-permanent wetland to an ephemeral wetland.

Last year, ANU analysis found Australia risked breaching the Ramsar pact due to convoluted water management and lack of obligation to meet and report on conservation goals.

In 2018, then water minister David Littleproud slashed the water recovery target in the northern basin by 70 gigalitres, putting further pressure on the marshes. The move was later described by the Murray-Darling Basin Royal Commission as 'deplorable' and 'unlawful'. In 2020 ICAC warned that the NSW government's water regulation had been 'overly favourable to irrigators' and that the Department of Primary Industries had focused on irrigators while restricting information available to environmental agencies.

Dr Celine Steinfeld, director of the Wentworth Group of Concerned Scientists, is out at the marshes. She's a critic of the 'credit' system used by the river operators, whereby inflows to the dam are predicted based on past rainfall charts and water is allocated to users before the rain actually falls.

In 2019, the predicted rain didn't come. The dam was at 4.7 per cent. In August, flows to the Macquarie River were shut off. In October, 3000 hectares of the marshes burned.

Steinfeld says the rainfall that year 'was worse than what was previously recorded', which foiled a system based on historical rain records. 'It all goes out the window with climate change,' she says.

It was the same year the Royal Commission warned that the plan ignored the potentially 'catastrophic' effects of climate change. The river was dry for over a year.

Garry Hall, a beef farmer whose family has run cattle on their property in the marshes for nearly 90 years, describes the credit model as 'archaic'.

He's also concerned that floodplain harvesting laws, which would regulate the amount of floodwater irrigators can divert when water flows onto their land, isn't going far enough to protect downstream flows.

'We need [floodplain water] licensed,' Garry says. 'But we've got to get it right. There's no good issuing licences that in three to five years, taxpayers have to go and buy back because there were mistakes made.'

Proposed floodplain harvesting laws have been blocked three times in the NSW senate over concerns they don't protect enough environmental water or downstream communities. But Environment Minister James Griffin signed off on water sharing plans that allow floodplain harvesting in two northern basin catchments, and the government plans to do the same in the Macquarie Valley in 2023.

In a statement, a spokesman for NSW water minister Kevin Anderson said, 'This reform will benefit water users, downstream communities, and the environment, with up to 100 billion litres of water to be returned to the floodplains in the northern river valleys each year.'

The government's own modelling shows the laws would only improve environmental water flows by 0.2 per cent each year in the Macquarie Valley.

'Who's in government for this next term will either secure the future of the wetlands and the downstream communities that depend on the river, or continue the disgraceful decline of the wetlands,' says Garry.

* *Model or monster?*, p. **99**
Where giants live, p. **187**
Talara'tingi, p. **193**

TRIALS OF THE HEART

Nicky Phillips

Around lunchtime on a warm March day in 1999, Kathleen Folbigg went to check on her sleeping 18-month-old daughter and found her pale and unresponsive. Folbigg, alone in her house in Singleton, Australia, called an ambulance while she tried her best to resuscitate the child. 'My baby's not breathing,' she said, pleading for them to hurry.

'I've had three SIDS deaths already,' she explained, referring to sudden infant death syndrome – a largely unexplained phenomenon that typically affects infants in their first year, as they sleep.

Around 9 pm that night, pathologist Allan Cala conducted an autopsy on the baby, named Laura, at the NSW Institute of Forensic Medicine in Glebe, a suburb of Sydney. In his report, he noted no evidence of injury and no medications, drugs or alcohol in her system. He mentioned some inflammation of the heart, possibly caused by a virus, but surmised that it could be incidental. Instead, Cala opined on the improbability of four children in the same family dying from SIDS. 'The possibility of multiple homicides in this family has not been excluded,' the report stated.

Four years later, in May 2003, a jury found Folbigg guilty of murdering three of her children – Patrick, Sarah and Laura – and of the manslaughter of her first son, Caleb. Because there were no physical signs of foul play in any of the deaths, the case had rested entirely on circumstantial evidence, including the unlikelihood of four unexplained deaths occurring in one household. Lightning doesn't strike the same person four times, the prosecutor told the jury.

Folbigg was sentenced to 40 years in prison, and became known as Australia's worst female serial killer. But in 2018, a group of scientists began gathering evidence that suggested another possibility for the deaths – that at least two of them were attributable to a genetic mutation that can affect heart function. A judicial inquiry in 2019 failed to reverse Folbigg's conviction, but in November 2022, the researchers will present a bolus of new evidence at a second inquiry, which could ultimately end in freedom for Folbigg after nearly 20 years behind bars. More than 90 scientists signed a March 2021 petition arguing for her release on the basis of that evidence.

The inquiry will have to grapple with how science weighs the evidence for genetic causes of disease, and how that fits with the legal system's concept of reasonable doubt. But it will have help. Thomas Bathurst, the retired judge leading the inquiry who will decide Folbigg's fate, has granted permission for the Australian Academy of Science in Canberra to act as an independent scientific adviser. The academy will recommend experts to give evidence, and will look at questions asked of those experts to ensure their scientific accuracy.

This will probably present the science more accurately than at the original trial, says Jason Chin, a legal academic at the University of Sydney who studies the way science is used in courts. And this case could have implications for how Australian legal proceedings consider scientific evidence in other cases, says Chin.

Sudden suspicion

Folbigg's four children died over a period of ten years. Caleb was just 19 days old in 1989 at the time of his death. Patrick and Sarah were eight and ten months old, respectively. Soon after Laura's death, Folbigg was placed under suspicion and eventually stood trial in a case that became a dramatic public spectacle. At the time, multiple SIDS deaths in a single family were viewed with suspicion, particularly against mothers.

That suspicion traces at least in part to Roy Meadow, a British paediatrician who studied child abuse. In 1997, he popularised the idea that 'one sudden infant death is a tragedy, two is suspicious and three is murder, unless proven otherwise'. At Folbigg's trial, this line of thinking clearly influenced some experts' testimony. And several leading pathologists testified that they had never come across multiple SIDS deaths in one family, implying that it was all but impossible – despite there being such cases in the literature, says Emma Cunliffe, a legal scholar at the University of British Columbia in Vancouver, Canada, who wrote a 2011 book about Folbigg's case called *Murder, Medicine and Motherhood*.

In fact, by the time the trial started, scientists had been expressing concern about Meadow's idea for a number of years, and particularly its use in legal cases. Sally Clark, a mother jailed under similar circumstances in the United Kingdom, had successfully challenged her conviction partly on the basis that Meadow's arguments were statistically unsound and made unsupported assumptions about the rate of SIDS. The idea was largely discredited by 2003, and Meadow was eventually struck from the UK medical register in 2005 because of misleading testimony he had offered during Clark's trial.

But Folbigg had other circumstantial evidence weighing against her, says Cunliffe. A selection of diary entries in which Folbigg expressed guilt and remorse about her deficiencies as a mother were presented by the prosecutor essentially as a confession. The prosecution's case relied heavily on the testimony of her husband, Craig Folbigg, who, on reading the diaries, became convinced that his wife had killed their children. The jury agreed. Several attempts to appeal the decision failed.

Folbigg has always maintained her innocence, however, and has not stopped fighting for her freedom. In August 2018, after a petition from her lawyers, the attorney general of New South Wales, Mark Speakman, announced that there would be a review into her convictions on the basis of fresh evidence around multiple cases of unexpected death in the same family. This would result in the first inquiry into the conviction.

Genetics enters the picture

As part of their preparations for the inquiry, Folbigg's lawyers approached Carola Vinuesa, a geneticist at the Australian National University (ANU) in Canberra at the time, to sequence and analyse Folbigg's DNA. The idea was to see whether she carried any mutations that, if inherited by the children, might offer an alternative explanation for how they died. Vinuesa agreed to help. Her colleague, Todor Arsov, a geneticist who lived in Sydney, travelled to the nearby Silverwater Women's Correctional Centre, where Folbigg was being held, to collect a sample from her.

That December, Arsov joined Vinuesa in her kitchen to scroll through sequence data looking for variants linked to sudden death. Within 20 minutes, they both came across something interesting: a variant in a gene called calmodulin 2 (*CALM2*).

Humans have three calmodulin genes, encoding identical proteins that bind to calcium and control its concentration in cells, which helps to regulate the heart's contractions, among other things. Mutations in these genes are extremely rare, but people who have them often have serious cardiac conditions; sudden deaths have been reported. Vinuesa thought the find was worth further investigation. She suggested to Folbigg's lawyers that they attempt to sequence DNA from the children and their father.

By this point, the inquiry was under way, and Vinuesa was asked to join a team of genetics and cardiology experts who would provide advice. In early 2019, the experts met. They included geneticist Matthew Cook, a colleague of Vinuesa's at ANU, who joined remotely, and three genetics specialists from the state's health department in Sydney. Jonathan Skinner, a paediatric cardiologist and cardiac electrophysiologist at Starship Children's Hospital in Auckland, New Zealand, also joined by video link.

Scientists had managed to obtain DNA from the four children. A hospital in Sydney had Sarah's frozen fibroblast cells from her autopsy in 1993, and the coroner's court in Glebe had frozen liver tissue from Patrick's 1991 autopsy. A laboratory in Melbourne sequenced a whole genome from one child's heel-prick samples, which are routinely collected at birth, and retrieved just the protein-coding portion of the genome from another's heel-prick card. Craig Folbigg declined to provide DNA.

By agreement, the experts separated into two groups to analyse the Folbigg genomes. Vinuesa and Cook, from Canberra, formed one. The experts from Sydney – genetic pathologist Michael Buckley, clinical geneticist Alison Colley and Edwin Kirk, who specialises in both fields – formed the other, with input from Skinner. (Members of both teams either did not respond to or declined requests for interviews with *Nature* to avoid jeopardising their involvement in the upcoming inquiry.)

To decide whether a variant causes disease, geneticists look at various lines of evidence. These include: whether a variant is rare or absent in the population, which would imply that it has been selected against; whether there are any clinical manifestations of disease in people or families who do carry the variant; and whether studies in cells or rodent models confirm that the variant can have an effect on protein function and health.

Each of these kinds of evidence can be used to score a variant using a five-tiered scale created by the American College of Medical Genetics and Genomics (ACMG), ranging from benign to pathogenic. In the middle of the scale are 'variants of uncertain significance', a messy classification that neither clears nor implicates a given mutation in causing disease.

If the inquiry found that a genetic variant in one or more of the children was 'likely pathogenic' or 'pathogenic', it could raise enough reasonable doubt about Folbigg's guilt to overturn her convictions, says Cunliffe.

Both teams started with a big picture approach, using bioinformatics tools to scan the genomes for rare variants. These alone are not a reason for concern or a sign of disease – most rare variants are harmless – but they're a good place to start when looking for the cause of a potential genetic condition. The Sydney team found 279 among the five Folbiggs. By filtering for genes linked to cardiac diseases, respiratory disorders or sudden death, they whittled the list down to just nine, each found in different combinations in the four children and their mother. Within that list was the variant in *CALM2* that Vinuesa had identified. It was present in Folbigg and her two daughters, Laura and Sarah.

Vinuesa and Cook in Canberra went through a similar process using their own bioinformatics tools. Overlap between the two teams' findings was good: both identified *CALM2* and a variant in the gene *MYH6*. The Canberra team also highlighted a mutation in a third gene, *IDS*, which is involved in a metabolic disease called Hunter syndrome. This can cause seizures and death, and was found in one of the boys, Patrick.

But when the teams issued their reports in March 2019, they differed in how they classified the rare variants. Although the Canberra team suggested that both the *CALM2* and *IDS* variants were 'likely pathogenic', the Sydney group concluded that none was.

In April, the hearings started. Over three gruelling days, genetics and cardiology experts were quizzed by the lawyer assisting the inquiry. The two teams ultimately agreed that the variants in *IDS* and *MYH6* were of uncertain significance and unlikely to be responsible for the children's deaths. But the two groups could not agree on how to classify the variant in *CALM2*.

Vinuesa and Arsov, who was involved in the Canberra team's report, told the hearing that the classification of 'likely pathogenic' was based on several ACMG criteria: the variant seemed to be completely new because it was absent in population databases at the time (it has since turned up on *CALM3* in one database). And computer simulations predicted it would be damaging to the function of the protein it encodes. The team also argued that there was a plausible pathway to explain how the variant triggers sudden deaths.

But members of the Sydney team challenged this classification, in part because Folbigg was apparently healthy even though she had the variant. Vinuesa argued that Folbigg's health status wasn't entirely known. She also explained that it was common for some people with a disease-causing variant not to show obvious signs or symptoms.

Skinner, the cardiologist, agreed that this was the case with some inherited heart conditions. He also agreed that mutations in calmodulin genes do produce life-threatening arrhythmias – irregular heart rhythms. But he told the inquiry that the arrhythmias 'don't seem to present in an infant who's quietly asleep'. Moreover, he added that in the literature, 'there's not a single case of a sudden death under the age of two'.

A disagreement unresolved

After the hearings had ended, but before the head of the inquiry, former judge Reginald Blanch, had released his findings, Vinuesa emailed several scientists who work on calmodulin, hoping they could help to settle the disagreements. Peter Schwartz, a cardiologist specialising in arrhythmias of genetic origin, wrote back almost immediately. Schwartz, at the Italian Auxological Institute in Milan, was on the team that had found a link between mutations in calmodulin and sudden death in childhood. In 2015, he had helped to establish a registry of people with known pathogenic mutations in the *CALM* genes, called the International Calmodulinopathy Registry.

Schwartz told Vinuesa that he and colleagues had just published a paper that mentions a family with a mutation at the same location as the Folbigg variant, but in a different version of the calmodulin gene, *CALM3*, and with a different amino acid substitution. In that family, a 4-year-old boy had died suddenly, and his 5-year-old sister had had a cardiac arrest but survived. Their mother, whose health seems unaffected, is mosaic for the variant, meaning that she carries it in only some of her cells. Vinuesa understood the significance – a new variant that results in an amino acid change at the same location of a *CALM* gene is considered strong evidence in the ACMG guidelines. In his letter to Vinuesa, Schwartz wrote: 'My conclusion is that the accusation of infanticide might have been premature and not correct.'

Vinuesa sent Schwartz's letter and the new paper to Blanch, and lawyers sent it on to the Sydney team. But the group of geneticists remained unconvinced. In their reply, they wrote that although the discovery of the similar mutation in another family 'has a significant impact on the likelihood' of its pathogenicity, that did not mean it caused the Folbigg girls' deaths, particularly in light of the fact that their mother is alive and seemingly healthy. For the variant to have caused the deaths of Sarah and Laura would require an 'exceptional clinical scenario', which is 'outside the range that has previously been reported in association with variants in this group of genes', the team wrote. 'Our classification of this variant remains that it is a variant of uncertain significance,' they concluded.

The disagreement between the two teams partly reflects the modern history of clinical genetics. The rapid pace of genomics research over the past two decades powered an exuberant search for pathogenic mutations. The medical literature is littered with papers that claimed to have identified dangerous gene variants that later turned out to be harmless. 'It became a bit of a wild west,' says Hugh Watkins, a cardiologist at the University of Oxford, UK, who studies genes that cause sudden cardiac death.

Scientists would link a gene to a certain disease because a newly found variant didn't show up in a small group of healthy people. Then they would use biochemical assays to show that the variant had some effect on the protein – not even necessarily linked to disease – and the paper would practically write itself. 'A lot of us in the field were sceptical that there just wasn't enough evidence, but these papers were quite easy to publish,' says Watkins.

The reckoning for this type of publication came roughly a decade ago, with the arrival of extremely large databases containing tens of thousands of genomes. Suddenly, it became clear that many supposedly deadly variants were actually relatively common in the population, and were therefore likely to be benign. 'At the time, we didn't imagine all the variation in our genome,' says Morten Salling Olesen, a molecular biologist specialising in genetic cardiac disorders at the University of Copenhagen. As a result, declaring a variant pathogenic now requires a fairly high bar of proof, says Watkins.

What sets the three calmodulin genes apart is that variations in them are very rare in population databases. Of all the highly conserved genes in evolution, 'they're a poster child', says Watkins. All vertebrates have exactly the same calmodulin sequences. 'That is evolution's way of saying they're important,' says Salling Olesen.

When scientists published details of the first mutation in 2012, it was eye-opening. 'People thought such mutations would probably be so severe that you would never have a live birth,' says Ivy Dick, an electrophysiologist at the University of Maryland School of Medicine in Baltimore.

In the past few years, the view has evolved again. In large population databases, a small number of seemingly healthy people have calmodulin mutations, albeit in different locations to known pathogenic variants. And it is possible for Folbigg to live with a pathogenic mutation while her daughters might have died from it. 'That's why it's so frickin' complicated,' says Walter Chazin, a structural biologist at Vanderbilt University in Nashville, Tennessee, who studies calmodulin.

In the end, it was left to Blanch to make a decision on the evidence presented by the Canberra and Sydney teams. In his report, released in July 2019, he wrote, 'I prefer the expertise and evidence of Professors Skinner and Kirk and Dr Buckley.' He concluded that he was in no doubt about Folbigg's guilt.

Vinuesa found this conclusion 'bewildering'. In a speech she later gave to a gathering of scientists and lawyers, she complained of the legal system's reliance on 'intuition' to inform decisions.

Schwartz, too, was perplexed. In a June 2021 letter to the president of the Australian Academy of Science, he wrote of his concern that Blanch and a subsequent appeal court had argued that Sarah and Laura Folbigg's deaths did not fit with what is reported in the literature. He countered that there were currently four cases in the calmodulin registry of sudden cardiac deaths or arrests in children under three while they were asleep. The two Folbigg girls 'reflect the natural variability associated with these conditions', he said.

The inquiry's struggle with scientific evidence did not surprise Cunliffe. When courts are presented with contradictory scientific perspectives, they tend to pick sides, preferring evidence from one expert, she says, and ignoring the uncertainty. Nevertheless, Cunliffe says she finds it 'astonishing' that scientists have a possible explanation for how the children died, and yet the court 'doubled down on Folbigg's guilt'.

Some of that undoubtedly comes down to the contents of Folbigg's diaries and the evidence she gave about them at the inquiry, which Blanch said strongly influenced his decision. For instance, in November 1997, two months after Laura was born, Folbigg wrote that she thought she handled Laura's crying better than she did Sarah's. 'With Sarah all I wanted was her to shut up. And one day she did,' she wrote. Another entry read: 'I feel like the worst mother on this earth. Scared that she'll leave me now. Like Sarah did. I knew I was short-tempered & cruel sometimes to her

& she left. With a bit of help.' Blanch concluded in his report that Folbigg's diaries were 'virtual admissions of guilt for the deaths of Sarah, Patrick and Caleb, and admissions that she appreciated she was at risk of causing similarly the death of Laura'.

A new inquiry

With the latest inquiry under way, Bathurst, who will preside over the proceedings, has stated that he intends to form his own opinion about the evidence. At the first hearings next week, he will consider the functional genetic findings gathered by scientists since the first inquiry, which they say demonstrates that the *CALM2* variant is pathogenic.

One of those researchers is Michael Toft Overgaard, a protein scientist at Aalborg University in Denmark, who was part of the team that discovered the first mutation in a calmodulin gene in 2012.

After the first inquiry, Vinuesa emailed Overgaard and asked whether he could perform a functional assay to determine the cellular effects of the Folbigg variant. Overgaard wasn't familiar with the case, but was drawn to the idea of seeing 'another piece in the puzzle to try and figure out how calmodulin works', he says.

Overgaard asked postdocs Helene Halkjær Jensen and Malene Brohus to do the lab work. Everything about the project was secretive: even other researchers in their lab didn't know about it. 'We had a folder on our computer called CSI,' says Jensen, a reference to the popular US television drama about crime scene investigators.

Jensen and Brohus spent weeks painstakingly making calmodulin proteins with the Folbigg mutation, known as G114R, in which the amino acid glycine (G) at the 114th position of the protein is replaced with an arginine (R). For comparison, they created proteins with two other calmodulin variants known to cause severe arrhythmias, G114W and N98S.

One of the first things they discovered was that the G114R variant cannot latch onto calcium effectively. The team thought this was important, because the effect was similar to that seen with the other two deadly variants. Further experiments revealed that G114R impairs how calmodulin attaches to two crucial channels that control the movement of calcium into the cell. Jensen says the results were compelling, but the team knew the most convincing evidence would be to show exactly how these channel impairments looked in a cell.

For that they asked Dick, who studies the protein CaV1.2, a channel that shepherds calcium into the cell. Calmodulin triggers this channel to close once enough calcium has entered. Overgaard asked Dick to look specifically at whether the mutation impaired the closure of CaV1.2. Dick had never heard of the Folbigg case, but to prevent her team from introducing any bias, she relabelled dishes of cells to hide their provenance. Sure enough, the Folbigg variant delayed the channel closure, letting extra calcium into the cell. 'That's what we know to be one of the signatures of a pathogenic calmodulin mutation,' Dick says.

But that wasn't the only effect the variant had. Wayne Chen, who studies calcium channels at the University of Calgary in Canada, was asked to conduct similar experiments on a ryanodine receptor, a channel that controls the release of calcium into the cell from intracellular stores. As it does with CaV1.2, calmodulin binds to ryanodine receptors and triggers the channels to close. This prompts the heart muscle to relax. When Chen's team expressed the G114R variant in human cells, that channel had trouble closing, too. The combined effect of the variant on both channels will increase calcium in the cell, says Dick, which increases the chance of arrhythmia. 'If you asked me, "Would this mutation be likely to cause sudden death?", I would say somebody with this mutation is at very high risk of that,' she says.

In November 2020, the international team published its findings in the journal *EP Europace*. The researchers concluded that the mutation that the girls carried set the stage for a fatal arrhythmic event that could have been triggered by an infection. Both girls reportedly had respiratory illnesses before their deaths, and the heart inflammation in Laura's autopsy now looked more relevant than ever.

Many scientists who spoke to *Nature* find the functional evidence persuasive. Watkins, for example, says the team used sophisticated assays that 'faithfully' reflect how the variant impairs crucial functions in the heart cell.

Salling Olesen cautions that functional effects don't necessarily equal disease. For instance, the Folbigg variant might have only a small effect on heart function, he says. 'It's hard to know.'

But Watkins says: 'With a variant that no one else has seen, I don't think it could be more convincing than what they've got.' He adds, 'The case that this is plausibly a cause of sudden death in early childhood is strong.'

Enter the academy

The 2020 study prompted the highly respected Australian Academy of Science to get involved. Its chief executive, Anna-Maria Arabia, was already interested in seeing the academy advocate for better representation of science in legal proceedings when she heard from Vinuesa after the first inquiry. Once the *EP Europace* paper was published, the academy and Folbigg's lawyers used it as the basis of a petition calling for the governor of New South Wales to pardon Folbigg. More than 90 prominent scientists signed the letter, sent on 2 March last year – among them Nobel laureates Elizabeth Blackburn and Peter Doherty, along with experts in paediatrics, cardiology and genetics. Arabia says not one person she approached knocked her back. 'That's how convincing this paper is,' she says.

Hearings for the inquiry will begin on 14 November 2022, and scientists and clinicians from around the world will be called to give their opinions on the evidence. Although the hearings will focus on the new genetic findings, experts will be called to discuss other aspects of the case, such as the boys' deaths. And, early next year, psychologists have been scheduled to give evidence about the diaries. Although the documents helped to secure Folbigg's conviction, some psychologists have challenged the way that they were interpreted.

The hearings will be a chance to test the academy's role as an independent scientific adviser. So far, commissioner Bathurst has given the academy a broad scope. Arabia says the academy will be giving advice on the most appropriate experts, ensuring questions asked by lawyers are scientifically accurate and that there are robust scientific discussions. 'That was missing in the last inquiry,' she says.

Beyond the Folbigg case, the involvement of the academy could set the stage for a new era of science in legal proceedings. A common criticism of the use of expert witnesses is that lawyers are ill-equipped to know the most appropriate people to call, something the academy hopes to change. 'It's an interesting sign of the growing openness to scientific expertise in the Australian court system,' says Cunliffe.

But Chin warns that the process could still be open to bias. Scientists 'have an interest in their theory being the dominant one', he says. Cunliffe, who thinks that Folbigg was wrongfully convicted, agrees that professional associations can be political, too, and so allowing them into the courtroom is unlikely to be a panacea. Still, she says, the academy's involvement 'is a very good thing for the Folbigg inquiry'.

The outcome of the inquiry won't be known for months. But if Bathurst finds that the genetics offer a reasonable explanation for the deaths of the Folbigg girls, it could eventually spell release for their mother, about five years before she would have been eligible for parole. Folbigg wrote in 2006 that she just wants the truth to be uncovered. 'That day I shall not gloat, or say, "I told you so" I'll simply cry and keep crying all the tears that are due to me.'

Postscript

On 5 June 2023, Kathleen Folbigg was unconditionally pardoned and released from jail, following the recommendation of the NSW Attorney-General Michael Daley. The recommendation came after the head of the inquiry into Folbigg's convictions, Thomas Bathurst, stated that he had reached 'a firm view that there was reasonable doubt as to the guilt of Ms Folbigg for each of the offences for which she was originally tried'.

✱ *Model or monster?*, p. **99**
 Do we understand the brain yet?, p. **262**

GALAXY IN THE DESERT

Jacinta Bowler

Standing in the West Australian desert, at the site of what will one day be the SKA-Low, is uncomfortably hot, but beautiful. The only smells are outback breeze and sun-baked plastic marquee.

Red dirt littered with scraggly trees and tufts of grass goes on as far as the eye can see; the scrub has flowers I've never seen before. But this place is almost overwhelming in its remoteness – there's no wi-fi, no phone signal – almost no way to contact the outside world.

This lack of communication is by design.

The place was picked specifically for its inaccessibility. It's four hours' drive to the nearest town, Geraldton, and eight hours to make it back to Perth. The Shire of Murchison in which it sits has no recognised towns, a population of around 100 people and is approximately the size of the Netherlands.

By the end of the decade it will be home to a sprawling 65-kilometre long forest of metal trees. This eerie plantation will be quieter than most – its job will be to listen out for faint radio signals from the earliest parts of the universe, and record 'leaking' radio waves from nearby alien worlds.

To listen to the sounds of the universe, phone signal, planes, wi-fi, even the reversing sensors on cars, will need to be silenced. If these antennas were ears, the radio signals blasting from our human-made devices would be the piercing audio feedback squeal at a rock concert.

The area I'm standing on – Inyarrimanha Ilgari Bundara ('sharing sky and stars' in the local Wajarri Yamaji language), the CSIRO Murchison Radio-astronomy Observatory – has very minimal infrastructure. This makes the site radio quiet – an area where radio transmissions are restricted to ensure that the telescopes can do their job. The telescopes are bold projects, billions of dollars in the making. The most powerful of their kind on Earth, they will be able to peer further into the universe than any other telescope before. They could tell us more about black holes, gravitational waves and almost every part of the universe – back to the earliest days after the Big Bang. They could even help us listen to aliens.

But far above is another technological wonder – mega-constellations of satellites, beaming down radio waves to provide internet to the world, including these radio-quiet areas. One or two satellites is manageable, but with businesses shooting thousands of satellites into space – around 1600 in 2021 alone – the sky is beginning to get noisy.

Connect with each other, or hear into the origins of the universe? Which is more important, and are both possible? Proponents of both are in an uneasy partnership to find a solution.

The SKA pipe dream

Conceived in 1991, the SKA project envisaged antennas spread over thousands of kilometres to 'simulate' a single device with the collecting area of a square kilometre. This engineering feat would have created undoubtedly the largest radio telescope of its kind on Earth.

The final design, which has just begun construction in both South Africa and Australia, is two distinct giant telescopes – the SKA-Low in Murchison, and SKA-Mid, whose core station is inside the Meerkat National Park in South Africa. The whole thing is run by an intergovernmental organisation called the SKA Observatory (SKAO).

When the SKA-Low is complete, around 2030, 131 072 total

antennas will form a 65-kilometre-wide spiral pattern, reminiscent of one of the galaxies it will be able to listen to. (The SKA-Mid looks a little more traditional, with 197 steerable dishes and 150 kilometres between the furthest two.)

Radio astronomy had humble beginnings. In 1939 amateur Grote Reber erected a 90-metre dish in his Illinois backyard and started surveying the sky.

Australia didn't get in on the act until 1945 during World War II, when scientists from the CSIRO and the RAAF used a radar station to observe radio waves coming from the Sun that had been messing with radio equipment. This sort of experiment could be done today with a TV antenna – if there weren't louder TV signals getting in the way.

Sixteen years later, Australia got its first big radio astronomy telescope – the 64-metre-diameter Parkes radio telescope, which received the live images of the Apollo 11 Moon landing in 1969.

Despite our slow start, radio telescopes are particularly well placed in the Southern Hemisphere. There are fewer people, and therefore less interference from human radio signals. But the Southern Hemisphere is also looking at a whole different patch of sky to scientists' northern colleagues.

If the Parkes telescope is movie famous because of the Australian classic *The Dish*, the movie *Contact* ensured the larger Arecibo Telescope's fame. Arecibo was a 305-metre-diameter radio telescope built into a natural sinkhole in Puerto Rico and was well known for programs to search for extraterrestrial intelligence. In 1974 the telescope was used to transmit a simple pixel picture message for aliens to nearby globular cluster Messier 13.

Compared to Arecibo and Parkes, the SKA (square kilometre array) telescopes are, technically, ginormous. SKA-Low's array forms a virtual radio telescope 65 kilometres in diameter, with a collecting area of more than 400000 square metres. The SKA-Mid acts like a 150-kilometre-diameter telescope.

Understanding the marvel of SKA-Low starts at the high-tech metal Christmas trees. At 2 metres tall, they have solid metal 'branches' near the top, which become larger wire branches below. These lower branches look just like wire coat hangers, but upside down, becoming larger as the tree gets closer to the ground.

Each tree works in a similar way to the antenna on your car, picking up radio waves. Except while your car radio might be trying to pick up the 'loud' FM signal band of 87.5–108 megahertz (MHz), the antenna is trying to pick up any faint signal from a wider range: between 50MHz and 350MHz. This includes everything from television and radio broadcasts to police scanners, CT scans and cordless phones. More importantly for the telescope, ancient hydrogen, galaxies and other space objects can also be mapped.

The antennas will be in groups of 256, each one called a station (although they look a bit like little forests). There are 512 stations – 131 072 total antennas. Some of these stations will cluster in the centre, while the rest arc out in three spiral arms. About 65 kilometres will separate the ends of the longest spiral arms – a 40-minute journey at highway speeds.

From the air, it will look a little like a glinting metal galaxy – perhaps like a Hubble or Webb image, with red shimmering in the background.

Just one Christmas tree antenna isn't much better than your TV antenna – it picks up whatever it hears, with no ability to distinguish where it's coming from. Instead, you need antennas spread widely to create a larger bucket to catch those faint radio waves.

A traditional radio telescope is made up of one big three-dimensional dish. This dish receives the radio waves then redirects them towards a receiver in the middle of the telescope, which is a point above and in the centre of the dish usually elevated by pieces of metal. These dishes can also be manoeuvred to face a desired direction for radio signals.

But SKA's antennae are fixed. Instead, they are digitally pointed – a technique called 'beam forming'. With thousands of antennas spread over 65 kilometres, radio signals from the same source will hit the antennae at different times. Imagine a sphere of radio waves colliding with the sphere of our Earth. A 'point' of the sphere is going to be first to touch us. In this example let's assume that it's at the centre of the SKA-Low formation. First, the central antennas pick up the radio waves. Then, as more of the sphere interacts, it looks like a ripple effect, spreading outwards towards the edges.

To make sense of this, the technology behind the telescope – the central signal processor – uses aperture synthesis to match up the radio waves. (Aperture synthesis was first formulated by Australian radio astronomers Ruby Payne-Scott and Joseph Pawsey in 1946.) This creates a 3D telescope, but the third dimension isn't depth – it's time.

And SKA has the potential to find aliens.

Human communications, broadcasts and later televisions and mobile phones have been gushing radio waves since we started radioing to ships a century ago. Some of these regular radio waves are blocked by the ionosphere, but some – especially TV broadcasting and those from satellites – can leak outside of our planet: Earth's technosignature.

If another alien civilisation in our cosmic neighbourhood was doing the same, the SKA is the first telescope that could potentially pick up other techno-signatures. When complete, each SKA is so sensitive it could detect a signal from a mobile phone on Mars – an average of 225 million kilometres away.

'It doesn't necessarily need to be beamed towards us or deliberate,' says Professor Cathryn Trott, SKA-Low's chief operations scientist.

But while some might be fascinated by the alien component, Trott is more excited by SKA's other important strand of research: identifying ancient hydrogen that spewed into our universe shortly after the Big Bang. This hydrogen could allow researchers to map what the universe looked like at its 'epoch of reionisation' over

13 billion years ago. This could help scientists understand when the first galaxies formed, and what else was happening in these first years of the universe.

This cosmic number of antennas, plus the interferometry and beamforming to make it work, requires thousands of processing modules worth of computing power from the central signal processor. But this pales in comparison to the mind-blowing amount of data that comes out of it.

Whenever they are running, both SKA telescopes will produce up to five terabytes per second of measurement data each: equivalent to downloading 200 HD movies every second.

This data travels through optical wires to the Pawsey Supercomputing Centre in Perth, almost 800 kilometres southwest, where the vast amount of material is analysed at a processing speed of around 135 petaflops – that's almost 100 000 times faster than a top-of-the-line smartphone.

Once the supercomputer is fully operational, it'll be 25 per cent faster than the current fastest supercomputer in the world.

While the SKA telescopes are still in their early stages, the excitement is palpable. There's so much out in the universe yet to be uncovered.

But the long-planned, multi-billion-dollar project only works if the telescopes aren't just overhearing our planetary chatter.

Satellites far above

Almost a lifetime ago, the first artificial satellite was catapulted to the skies. The Soviet Union's Sputnik 1 was launched from Kazakhstan in October 1957 and for 21 days it transmitted a single repetitive beep tone that could be heard by any curious amateur radio operators.

Since then, we've been slowly adding satellites to our low earth orbit, launching around 100–200 objects into space every year.

Having eyes far above the Earth's surface is critical for science. There's the International Space Station, along with weather-, fire- and greenhouse gas-monitoring satellites. Satellite phones and internet have long been a mainstay of rural life.

When the Hubble Space Telescope launched into low Earth orbit in 1990, there were about 400 active satellites buzzing around.

By 2000 there were 700. Ten years later that number had just scratched 1000. But in the mid-2010s launch numbers began to skyrocket. By the time you read this, there will be over 10 000 satellites in Earth's orbit, around half of them active.

This explosion of satellites has been spurred on by two technological advances. Although Elon Musk's SpaceX wasn't the first to design reusable spacecraft or launch components (NASA's Space Shuttle was being reused back in the 1980s), the company has recently succeeded in creating partially reusable launch systems, pushing down the costs of rocket launches.

Smaller, cheaper components have also allowed more institutions and companies to launch satellites for a fraction of the previous cost.

This, along with the drive to give rural and remote areas internet access, has led to 'megaconstellations' of satellites. Starlink has launched more than 3000 of a planned 12 000-strong satellite fleet. Amazon has 3000 satellites planned and OneWeb is deploying another 600.

The more satellites there are, the more of the planet gets access to regular and reliable internet. The more companies, the more competition – potentially pushing down the price.

But for astronomers, according to Dr Phil Diamond, the SKAO director-general, 'they're a pain in the arse'.

Federico Di Vruno has broad shoulders, a bald head and eyes that crinkle up when he smiles. Originally from Argentina but now living in the UK, he spent our Zoom call sipping on maté – out of a silver curly straw.

He lives on the other side of the world to both SKA-Low and -Mid, and yet he's a key to the multi-billion-dollar projects being a success.

The spectrum manager tells me that he used to be a satellite engineer; he's still an engineer at heart.

'It gives me a really interesting perspective, because I understand satellite people,' he says.

'These constellation operators are engineers. Of course, they care a lot about astronomy, and they want to try to use space in a sustainable way.'

In 2018, when SpaceX launched the first few Starlink satellites, astronomers knew that these satellite constellations could be problematic.

And because Starlink's job is to provide satellite internet access around the planet, even remote or radio-quiet areas like Inyarrimanha Ilgari Bundara and Meerkat National Park and their highly sensitive radio telescopes are at increased risk from stray internet radio beams.

Generally, the frequencies of radio waves are carved out for different purposes, including broadcasting, radar, or mobile phones.

Even outside radio-quiet zones, radio astronomy has teeny slivers cut out of the allocations to allow for scanning the heavens. Today's radio astronomy needs access to the full spectrum of frequencies – galaxies have a different frequency to pulsars, which have a different frequency to ancient hydrogen. To do full sky surveys you need to listen across multiple wavelengths to get a proper picture.

This is why radio-quiet zones are so important. 'This unprecedented level of radio frequency interference control was a significant factor in Australia's successful bid to host SKA-Low,' writes CSIRO radiocommunications engineer Carol Wilson. 'It has already led to world-class astronomy results in frequency bands that cannot be used elsewhere in the world, particularly in 700–1000 MHz.'

But zone regulations are only enforceable on the ground, not in the air or in space. Satellites can legally do whatever their owners like.

SKAO's analysis suggests that without mitigation, SKA-Mid is likely to lose data between 10.9 and 12.75 GHz – the radio wavelength that satellites use to contact the ground. But with hundreds of thousands of satellites constantly beaming down, the Band 5b receivers – which range from 8 to 15 GHz – could be completely lost.

SKA-Low – which is observing in a lower frequency range – will have fewer issues. However, CSIRO is still studying the impact of having the satellite internet user terminals near the site.

Di Vruno's job is to work with satellite companies to come to an agreement on how megaconstellations and radio telescopes can exist in harmony.

'Lower Earth orbit is big but not huge – and definitely not infinite,' he says. 'The numbers [of satellites] we are managing right now: they're large, but not so scary.'

'But the plans out there to launch all these satellite constellations is quite scary because the numbers get up to 500 000 satellites.' In true engineer understatement, he adds: 'That's a lot.'

This has led to virtual meetings and straight talking with the world's largest tech and space companies. It has taken him to the United Nations Committee on the Peaceful Uses of Outer Space (COPUOS) to advocate on behalf of ground-based telescopes worldwide for access to 'dark and quiet' skies.

The February meeting of the COPUOS Scientific and Technical Subcommittee was the second time that dark and quiet sky protection was a distinct agenda item – Di Vruno and the team were ecstatic about the result.

'We have over 160 members now, and the four hubs are

working in many different work packages like coordinating satellite observations, engaging with industry and others,' he says.

'It's interesting to see how things have changed [since] we started having this conversation.'

Sharing sky and stars

Astronomers themselves are beginning to come up with answers. The team behind an SKA 'precursor' called the Murchison Widefield Array (MWA) have found a way to cancel out radio interference.

'We're trying to do radio astronomy here on the MWA with satellites creating all sorts of interference,' says Professor Elanor Huntington, a CSIRO executive director. 'The people who were doing that had to figure out a bunch of signal processing, to essentially look past all of the noise that the satellites were making.'

Using the MWA as a passive radar, 'they have now also figured out how to flip that round, so that they can make a company that is ... specially designed to look for those satellites for the space industry'.

Coordinating where satellites are allows researchers to schedule observations around them – but more satellites means more cuts into observing time.

Industry is also engaged. Despite a difficult start – Elon Musk allegedly told astronomers that if they were clever, they would just put their telescopes into space – Starlink has been working to try and solve both the brightness and radio problems.

'Starlink is actively avoiding some radio observatories in the world,' says Di Vruno. 'That's something that is being implemented now.'

Murchison's radio-quiet zone is a 260-kilometre-radius circle, and when Starlink launched in Australia, they left a 70-kilometre-radius hole over the most crucial inner area closest to the telescope.

Like many truces, this one is on uneasy territory. Recently, after assessing that the telescopes in the area use different frequency bands to the ones the satellites are using, that hole on the Starlink map quietly closed. And Starlink isn't just one satellite comms enterprise in the market. Di Vruno's bridge-building work is far from over. He's upbeat about his mission, although you can tell that there's tension there too.

'It is a challenge for sure. But I don't think that this is the end of radio astronomy at all.'

This much is certain: the SKA telescopes are going ahead, satellite constellations or not, and its proponents and creators are consumed with the possibilities it brings to astronomical discovery.

'I was in CSIRO when the first science data started coming out of [SKA predecessor] ASKAP. Going to the meetings and morning teas with the astronomers as they started bringing the first images … that's amazing,' says SKA-Low director Dr Sarah Pearce.

'I must admit to having watched the press conference for the first James Webb Space Telescope images and thinking how exciting it's going to be when we can do the same.'

✱   *A universe seen by Webb*, p. **78**
    Space cowboys, p. **183**
    Dark skies, p. **245**

A CITY OF ISLANDS

Helen Sullivan

The ancient city of Nan Madol, which translates as 'the space between', was built sometime between 700 and 1200 CE, on a coral reef just off the island of Pohnpei, in the Federated States of Micronesia. It was a city of islands, a miniature Micronesia consisting of 92 man-made bits of land. Micronesia itself is made up of 607 islands, of which the city of Palikir on Pohnpei is the capital. There is only one way for civilians to get to Pohnpei and most of the other islands in Micronesia. United Airlines Flight 154 takes off from Honolulu and, over the course of 22 hours, stops at Majuro and Kwajalein in the Marshall Islands, then at Kosrae, Pohnpei and Chuuk in Micronesia, before reaching its destination, Guam.

Standing in front of me at the check-in queue in Honolulu was an American bound for the US military base on Kwajalein. I asked him what he did there and he told me he was a military contractor and ex-soldier, currently employed at a 'facility'. This sounded mysterious, so I asked him what went on at the 'facility'. He lowered his voice: 'Oh you know, they do things and stuff.' When we began talking to the woman behind us, who was joining her husband, a soldier, on Kwajalein, it emerged that he was a mechanic, in charge of maintaining the island's rubbish trucks.

Only a thousand people live on Kwajalein, most of them American. Fifteen thousand Marshallese live on its closest neighbour, Ebeye, an island one-tenth of the size and one of the most densely populated places on earth. Though it is part of the

Marshall Islands, Marshallese may only live on Kwajalein with the permission of the US army. The relationship between Kwajalein and Ebeye mirrors South Africa's apartheid pass system: Ebeye's residents are allowed onto Kwajalein only to work and must leave every day after their shift ends.

Few of them chose to settle on Ebeye. Most are refugees, or the children and grandchildren of refugees, from the US nuclear tests on Bikini Atoll, carried out between 1946 and 1958. After World War II the Marshall Islands and Micronesia were made part of a single territory, the United Nations Trust Territory of the Pacific, and this group of more than 2000 islands was put under American control. The US decided to use this new territory to test nuclear weapons: the combined energy yield of the tests exceeded 7000 Hiroshimas. In the aftermath of Castle Bravo, a 15-megaton hydrogen bomb that was exploded over Bikini in 1954, John Anjain, a magistrate from nearby Rongelap, reported that 'women gave birth to creatures that did not resemble human beings: some of the creatures looked like monkeys, some like octopi, some like bunches of grapes.'

A little north of the equator, and just west of the International Dateline, Micronesia sits between China, the US and Australia. Pohnpei is 5000 kilometres from Hawaii, 4000 kilometres from Taipei and 3000 kilometres from the Australian city of Cairns. The UN created 11 trust territories around the world, but because of its location, the Pacific Trust Territory was the only one whose future had to be determined by the Security Council, rather than the General Assembly.

In 1986, the US terminated the trusteeship and Micronesia became independent. It had by then entered into a Compact of Free Association (COFA) with the US, which is still in place today. Its most recent version gives Micronesians the right to study, work and live in the US and provides Micronesia with financial and military support. The agreement is set to expire next year, a fact that has become newly significant as China tries to gain a foothold in the region. According to a recent report by the US Institute of Peace, 'perhaps to a greater extent than any other geographic area, the Pacific Islands offer China a low-investment, high-reward opportunity to score symbolic, strategic and tactical victories in pursuit of its global agenda.'

The US has never tested bombs in what became Micronesia. It hasn't been allowed to build a military base there either. Instead, it has recruited Micronesians to fight in its wars: per capita, Micronesians sign up to the US military at a higher rate than people in any US state. One in a hundred Micronesians is currently enlisted. They also died in greater numbers in Iraq and Afghanistan, per capita, than recruits from any US state, according to the *Christian Science Monitor*, which reported in 2010 that Micronesia had 'lost soldiers at a rate five times the US average'.

Photographs of ten of the dead hang on the wall of the arrivals lounge at Pohnpei Airport, where Kukulunn Galen, a representative from the tourism department, was waiting to meet me. I was the first journalist to visit since Micronesia's borders had reopened after Covid, she told me, putting a *wairenleng* of fresh flowers on my head. We looked at the portraits. 'This is one of the benefits of our relationship with the US,' she said, then laughed, realising what she had implied.

Walking to the centre of Kolonia, Pohnpei's main town, I passed sleeping dogs, small shops selling snacks, a UNICEF office in a shipping container, a post office, a satellite dish as big as the post office, the bright green tourism office with a sign showing the distance from Pohnpei to various cities around the world, and a billboard displaying two thermometers, which measured the island's Covid vaccination progress (70 per cent of the population). At the Red Cross office I met Diaz Joseph, a former chairman of the local branch, who now teaches at the college in Kolonia. I asked him where his brightest students end up. 'The military,' he said. 'It's unfortunate.' The US military holds many rounds of aptitude tests each year. There are few other means available to students who want to study overseas, and even then the army pays only for approved courses. Enlisting also means your flight off the island is paid for – a serious consideration for many young people. A one-way ticket costs more than $1000, a quarter of the average income on the island.

Once in the US, Micronesians have full working rights. But raising the money to get off the island without joining the army is difficult. First, you might turn to civil servants, who earn good salaries, for help. If that fails, and it usually does, your family will hold a raffle, gathering donations of coconuts, taro, bananas, rice, chickens and maybe a pig, selling tickets at 25 cents each. The winner gets the whole lot, Joseph told me, making the shape of a large pile with his hands.

There are more than 100 000 Micronesian citizens, but less than two-thirds of them live in Micronesia itself; nearly all of the others live in the US. Micronesia is the world's sixth smallest economy, with a GDP of just $400 million a year. A quarter of this comes from the US government under the terms of the compact; just under a quarter comes from selling fishing rights to vessels from other countries – mainly China, Taiwan and Japan. The hundreds of islands don't add up to much: together, they make up an area smaller than New York City. But the area between them is vast. Micronesia controls the world's fourteenth-largest exclusive economic zone – around 2.6 million square kilometres of ocean to which it has sole exploration rights. This is three times larger than China's undisputed zone, and twice as large as China's and Taiwan's combined. Chinese funds paid for the building that houses Pohnpei's Western and Central Pacific Fisheries Commission.

It's projected that in seven years' time, the waters of Micronesia will be so acidic that the coral reefs – which not only absorb carbon dioxide and support the fish on which its inhabitants depend, but are a buffer against rough seas and high tides – may no longer be able to grow more than they shrink, as the acid dissolves them. They are expected to stop growing altogether before long:

the Pacific Ocean has become 30 per cent more acidic over the last 200 years. The acidity is likely to increase by a further 150 per cent by the end of this century, making the ocean unliveable for many life forms.

Drought and rising sea levels threaten subsistence crops. Because Pohnpei is mountainous, it is relatively well protected. But on the smaller islands, coconut trees are being washed away and taro patches are inundated by salt water. Many local legends centre on food, and people on the island used gastronomical metaphors to explain things to me. ('These are my banana trees! These are my yams!' Joseph said of his students.) The stories about Nan Madol focus on food, and the cruelty and greed of the Saudeleur emperors, two brothers, Olosihpa and Olosohpa, who sailed to Pohnpei 'from the western sky'. The Saudeleurs were all-seeing and all-knowing and had many rules about who could eat what. Pohnpeians, who lived on sea snails, oysters, clams, fish, coconut and breadfruit, were not allowed to eat even a single louse; any louse they found had to be taken to the Saudeleurs. The emperors' dog, Watchman of the Land, made sure of their obedience. In these legends, Pohnpeians succeed at impossible tasks – obtaining a scale from a fish nobody has ever seen or, with the assistance of spirits, a feather from a mythical bird – assigned as punishment for even the smallest transgression (usually having eaten something they shouldn't).

Each island in Nan Madol had a specific purpose. There was an island for clam aquaculture, another for turtle husbandry, one for burying priests and another for punishing, torturing or executing people who failed to pay sufficient tribute to their rulers. The city, abandoned for more than 400 years, is now preserved as a Unesco World Heritage site. Pohnpeians managed to mine, move, lift and lower into place the basalt that built it. The weight of the stone would certainly sink a canoe – some pieces are estimated to weigh around 50 tonnes. Local legend attributes its building to magic. The extensive UNESCO documentation says only that building the city required transporting and placing 'an estimated two thousand tons of volcanic rock every year for at least three to four centuries without the benefit of pulleys, levels, metal tools or wheels'.

The Saudeleurs worshipped a god called Nahnisohnsapw. Nahnisohnsapw's medium on earth was a saltwater eel, Nahn Samwohl (a very eel-like name), who was 'large, foreign, frightening and ravenous', according to the historian David Hanlon. Nahn Samwohl was the evil version of the smaller, friendlier-looking freshwater eels worshipped by Pohnpeians. He lived on one of the islands of Nan Madol in a well, connected to other islands by underwater tunnels. Twice a year, in a ceremony marking the transition between *rak*, the season of plenty, and *isol*, the season of scarcity, he was fed turtle stomach (the two sons of a turtle had once expressed the desire to eat dog, another forbidden food).

The Saudeleurs were invaders; according to legend, until their arrival Pohnpei had been ruled by the descendants of the nine women and seven men who first discovered the island, guided there by divine winds and an octopus. When they arrived on the island, they built an altar. Pohnpei means 'on a stone altar'.

The Pohnpeians were liberated from Saudeleur rule by the son of the Thunder God, who had escaped imprisonment on Nan Madol and sailed to a mythical island where he impregnated a woman by squeezing lime juice into her eyes. Their son, Isokelekel, grew up hearing stories about the Saudeleurs and, sometime in the early 1600s, sailed to Pohnpei with 333 warriors. Isokelekel is still seen as the hero who liberated Pohnpeians from tyrannical – and centralised – rule. Isokelekel introduced the traditional system of Nahnmwarki, or multiple chiefs, which continues to this day, unbroken even by colonial rule. In 1886, Pohnpei was colonised by Spain. Germany bought it in 1898 and Japan claimed it after World War I, retaining it as a territory until 1945 – small, rusty Japanese tanks can still be found in its forests.

Augustine Kohler, the acting director in charge of Nan Madol, believes it was a place for sailors to trade, restock and pray for their journey ahead. He started his story of the ancient city by saying: 'This was the only time in Pohnpei's history that we were ruled by one person.' Nan Madol is a reminder of the dangers of falling under the spell of a foreign power. This is especially relevant at the moment, as Micronesia tries to maintain ties with both the US and China, using China's interest in the region as leverage in negotiations with the US (particularly when it comes to climate change adaptation).

And China is extremely interested. Between a pretty Catholic church and a baseball pitch in Kolonia is a large basketball court on which the words 'KOLONIA-CHINA FRIENDSHIP CENTRE' are painted in person-sized letters. It was built in 2017 in front of – and dwarfing – a small Japanese statue of Buddha. As we stood looking at it, Jasmine Remoket, a guide from the tourist office, smiled. 'If China sponsors something you will know,' she said, sweeping her hands across the sky.

Earlier this year China signed a security pact with the Solomon Islands, a leaked draft of which mooted the possibility of China establishing its first military base in the Pacific, less than 2000 kilometres from Australia. Shortly afterwards, China circulated a draft policing, security and data agreement with ten Pacific nations, prompting the Micronesian president, David Panuelo, to write to the other Pacific leaders warning that signing the deal would increase the likelihood of 'a new Cold War era at best, and a World War at worst', and was a distraction in the face of what he called 'our ceaseless and accurate howls that climate change represents the single-most existential security threat to our islands'.

Micronesia's seas are rising by more than a centimetre every year, three times higher than the global average. There are few places on earth threatened by climate change to the same extent as the low-lying Pacific islands. Twelve of the coral reef islands off Pohnpei have shrunk or disappeared totally in the past decade according to Patrick D Nunn, a professor of geography and sustainability at the University of the Sunshine Coast in Australia. Two more are likely to go over the next few years.

Among the recipients of Panuelo's letter were the four Pacific nations that recognise Taiwan: Nauru, the Marshall Islands, Palau and Tuvalu. These tiny states are subject to intense lobbying by both sides. The Solomon Islands switched allegiance from Taiwan in 2019, while Nauru has switched and switched back again in recent years. 'A war for Taiwan is equivalent to a war between China and the United States,' Panuelo wrote. 'Whoever wins in such a conflict, we will once again be the collateral damage.'

The countries collectively rejected China's proposal and in August, President Joe Biden announced that the 'first ever US–Pacific Island Country Summit' would be held the next month – an example of what Patricia O'Brien, who teaches Pacific studies at Georgetown University, called 'warp speed' diplomacy. This time, the leaders did sign a declaration. It promised $810 million in aid over the next decade, not including the additional funding expected when COFA is renewed later this year. (The Solomon Islands signed the declaration only after all 'indirect references' to China had been removed.) 'China did us a huge favour,' Bill Jaynes told me. Jaynes is American but has lived on Pohnpei for three decades; he is currently the only journalist on the island's only newspaper, the *Kaselehlie Press*. 'I've worked with many ambassadors who have said: "There will be absolutely no extension of the financial terms [of COFA]." In the last year and a half, that changed.'

I left for Nan Madol with Paul David, another guide from the tourist office. We drove out of Kolonia along the island's ring road – the complete circuit takes three hours – much of which is lined with single-storey houses, some painted in bright colours and all with a garden. Taro, banana and breadfruit grow alongside flowers – hibiscus, orchids, frangipani – and many houses have pig pens. Just beyond a sign saying 'Welcome to the Kingdom of U' (U is the name of the municipal area), the vegetation opened up to reveal a small bay. We turned left, drove down a hill, and entered an underworld: the roots of the mangrove forest.

Waiting in a motorboat was Ayler Gallen, a Nahn Saum, or traditional leader. Lying in the boat was his sleepy four-year-old grandson, Jose. We sputtered along a mangrove canal, out to the open sea. Mangroves grow on, in or near almost every structure on Nan Madol. We passed a wall that dropped to a small foothold of land at the edge of the water: this was Takai en Rihp Kapehd, 'the rock of the tightening stomach'. Pregnant women used to walk from one end to the other: if they managed to touch the far side, their child would be a warrior.

We passed into a wide canal, moving between various large basalt islands that sometimes could be glimpsed only in gaps between the monstrous mangrove roots. Then we began to slow down, and could no longer see the open ocean. To our right stood Nandowas, one of the most impressive islands of Nan Madol, surrounded by walls two storeys high. Its name means 'in the mouth of the high chief'.

'No one knew what was in the mouth of the high chief,' the Pohnpeian oral historian Masao Hadley writes in *Nan Madol: Spaces on the Reef of Heaven*. 'No one knew what he did inside. No one understood what was inside Nandowas. Nandowas was a place of war.' We got off the boat and I put my hand on one of the enormous columns. The front walls formed an entrance through which another, shorter set of walls guarded a central building. Plants grew everywhere: vines, trees, taro plants with leaves as big as a person, palms and ferns of every description. Trees pose the greatest threat to the man-made city – falling branches and growing roots are breaking up the islet walls and foundations while mangroves close off the canals – and are weakening its defences against rising seas and stronger storms.

That day, a group of workers cutting back mangroves had cleared a waterway thought to have been impassable for 50 years. This was part of a conservation project funded by the US State Department and due to end this year. Progress had been slow – fearful of spirits, the workers wouldn't stay after 3 pm. We followed the newly cleared route. Deeper in the forest, the only light came through small gaps in the thick canopy. David spotted a ray in the water, causing Jose to forget his seasickness and clamber to the front of the boat to see it. Gallen said that it was a blessing and would guide us back to open water. David sounded sceptical as he translated, but the ray continued to swim slightly ahead of the boat until the canal opened into a broad waterway. Then the ray slowed down, we passed it, and it was gone.

With David as translator, I asked Gallen how he had felt travelling along that canal for the first time in 50 years. 'He feels sad when he looks at it because it is so different from when he was small. He says that back then it was well maintained, people really looked after it. He is happy to see that it is being cleared now, but the project will end in January and he knows the work will not be done by then.'

On the way home, David bought a bottle of sakau and dispensed a little to each of us in empty water bottles. Sakau is made by pounding kava root and mixing it with hibiscus sap; it tastes like wheatgrass, but muddier. It is meant to cause a feeling of wellbeing and calm: Pohnpeians drink it ceremonially and for fun, and before discussing difficult subjects.

David told me about his nephew, a straight-A student who had hoped to go to medical school. His father, David's brother, had died around the time the boy was applying to college, and when four weeks passed without a reply he went to the US embassy to sign up with the army. David had urged him not to go. 'When he arrived home from the embassy, his acceptance letter for medical school was waiting for him,' David said. But his nephew felt it was too late. He did a tour in Afghanistan and now lives in the US. David's son is finishing high school and he, too, wants to join the military. David and his wife are pleading with him not to.

✱ *A subantarctic sentinel*, p. **85**
     The Torres Strait Islander elders lawyering up to stop their
       homes from sinking, p. **142**

EARS

Heather Taylor-Johnson

Most ear issues lean toward vertigo and hearing loss, and doctors will not diagnose a patient with Ménière's disease unless they experience both. Before my diagnosis, I thought hearing loss meant gaining silence. Since my diagnosis, I've gained noise.

It's my cochlea in the inner ear, shaped like a common spirula shell or a Cuban brown snail shell. The cochlea's function is to change the vibrations from the middle ear into nerve signals that travel to the brain so that we can hear. My ear falters there, in the tiny hairs of the cochlea, and the noise I hear every day sounds as if I were holding up a gigantic shell to my ear: a queen conch or a Triton's trumpet. You may imagine I'm René Magritte's *Shell in the form of an ear* (*Untitled*), a surrealist painting of a gigantic ear-shell shaped like a Triton's trumpet. Alone on a beach with its impossible size, the ear-shell is completely out of place, so cavernous. I am bereft at the enormity of its sucking-in of all that air, ripe with an uproarious swish of seagulls, which are clearly there but out of frame. When I look at the painting I can hear the absent presence of the seagulls, and they're loud.

There'd be no noise to contend with on the moon because we'd all be profoundly deaf. You can't hear on the moon, the atmosphere's all wrong. To hear, we need the vibrations of Earth's air, its cycles of condensation and rarefaction. Each cycle is called a Hertz, the lowest note on a piano being 4186 Hz while the highest is 27. Most human capacity for hearing lies between 20 to 20 000 Hz so most people can hear a piano's scale, though on the moon, pianos would be obsolete. I would miss them. I would miss music. But there's this: if every survivor of our Earth's apocalypse made it to the moon, we'd all go deaf together, which means we'd all have to learn sign language, and maybe that would make up for the deprivation of music because instead of dancing to music, we'd be dancing to a silent language, and wouldn't that be beautiful!

In Sean Williams's book *Impossible Music*, a young man named Simon wakes up in the middle of the night to find he can't hear a thing: 'Silence has pressure and weight. It grinds you down, and I suppose that could have been what woke me.' He's a musician and composer for whom music is identity-making and confirming. It's his past, his present and was going to be his future. In trying to bring silence and music together, in trying to keep his identity from completely shattering alongside his hearing, he refers to musical theorist John Cage, who said, 'Every something is an echo of nothing.' Simon asks, 'What if nothing is the whole point, and the something only gets in the way?'

In an address to the convention of the Music Teachers National Association in 1957, John Cage also said this:

> There is no such thing as an empty space or an empty
> time. There is always something to see, something to hear.
> In fact, try as we may to make a silence, we cannot. For
> certain engineering purposes, it is desirable to have as silent
> a situation as possible. Such a room is called an anechoic
> chamber, its six walls made of special material, a room
> without echoes. I entered one at Harvard University several
> years ago and heard two sounds, one high and one low. When
> I described them to the engineer in charge, he informed
> me that the high one was my nervous system in operation,

the low one my blood in circulation. Until I die there will be sounds. And they will continue following my death. One need not fear about the future of music.

Music as silence, silence as music.

I can't relate to Simon from *Impossible Music*. I don't understand silence. Rather I identify with his girlfriend G, who has a vexing form of tinnitus that's resulted in extreme hearing loss. It's not only about phantom noises whirring on and on, but also an inability to compartmentalise hearing, an inability to filter out background noise and focus on the foreground. It's a cluttered cacophony that fills a space so demandingly one gets used to nodding, as if they can catch the words flung in their direction when really there's no hope at all. At parties, pubs and restaurants it's unfeasible someone with this type of hearing loss will ever be fully invested in any conversation. The painter David Hockney called it 'noise pollution' and it drove him away from art exhibitions, though on the flip side he claimed his hearing loss made his painting sharper: 'I think my going deaf increased my spatial sense, because I can't get the direction of sound. I feel that I see space very clearly, and that's because I can't hear it.' This led to vivid English landscapes with the same distinctive, almost animative style of his Los Angeles swimming pool paintings of the '60s and '70s, but these had a Van Gogh feel. Hence the Hockney–Van Gogh exhibition in 2019 called *Two Painters, One Love*. Noise pollution can result in exquisite works of art.

Beyond exhibition openings, parties, pubs and restaurants to remote areas littered with multiple slow-turning blades of wind turbines, noise pollution is everywhere. It even infiltrates our oceans. In the underwater world, right whales are losing their hearing because of it and can no longer communicate as they once could with other right whales, which means they can't find each other so easily. More and more they're spending time on their own.

In 1929 René Magritte moved to France, where he met his great friend and rival Salvador Dalí. Together they would define Surrealism by painting interpretations of their dreams. (Magritte said, 'If the dream is a translation of waking life, waking life is also a translation of the dream'.)

After meeting Pope Pius XII at a private gathering, Dalí declared himself a Catholic and began experimenting with the image of the Virgin Mary in his paintings. Two of those paintings strike me deeply and I consider them companions: *Madonna* – a rendition of Raphael's *Sistine Madonna* situated in a gigantic ear – and the beguilingly titled *Van Gogh's ear cut dematerializing from its frightening existentialism and exploding in the manner of a pion during the dazzling of Raphael's Sistine Madonna* (sometimes shortened to *Cosmic Madonna*), in which another rendition of the *Sistine Madonna* is key. Was Catholicism, which he'd rejected for the first half of his life, whispering in his ear? Or was it the madness of Vincent van Gogh? Or was it the notion of 'ear'?

Some might say the myth of Van Gogh's severed ear is greater than his art, though I am not one. *Sugababe* is a work in which Van Gogh's ear has become a work of art. Dutch artist Diemut Strebe began with a 3D bioprinting in the shape and size of Van Gogh's ear, using a dissolvable sugar-polymer scaffolding which she injected with the living cartilage cells of Lieuwe van Gogh, the great-great-grandson of Vincent's brother Theo. For several weeks the ear nested inside a bioreactor, and eventually the scaffolding dissolved while the cells thrived. The ear, with one-sixteenth of the same genes as Vincent van Gogh and living in a nutrient-rich solution, is alive.

One hundred and twenty-five years after the artist severed his left ear, Strebe has grown it back, and it can hear. If you speak to it through a microphone, the sound will travel to a computer processor that simulates nerve impulses. Noam Chomsky was the first person to speak to the ear, and when he stopped talking, the ear crackled back at him, the feedback being a response to silence, not words. Chomsky was therefore hearing an ear hearing silence, which goes to show silence is never silence.

By the age of 12, lover of musical instruments Evelyn Glennie had become profoundly deaf, so she took up percussion. With drums, she could still hear the music because she could *feel* the music. Incredibly gifted and determined, Glennie became the first person to successfully establish and maintain a full-time career as a solo percussionist. 'My career and my life have been about listening in the deepest possible sense,' she writes. 'Losing my hearing meant learning how to listen differently, to discover features of sound I hadn't realized existed. Losing my hearing made me a better listener.'

Glennie performed on the soundtrack of *The Sound of Metal* – a stark and poignant film about a heavy metal drummer losing his hearing – and interviewed the film's star, Riz Ahmed, for *The Evelyn Glennie Podcast*. They talked drumming and listening and Deaf culture, and so often the three were inseparable. When she asked him if he'd learned anything that he'll take away with him, that's transformed him in any way, his answer was thus:

> I feel the Deaf community taught me the true meaning of listening. You know, listening is so different from hearing. Hearing is a physical sense; it's a biological process. Listening is something to do with your attention and your energy, and where you place it, and how you hold space for someone else's energy with your attention. It's much more – it's an act of

absorption, of osmosis. It's a meditative act ... at times a kind of spiritual act ... it's about ... removing a barrier between yourself and the other, I think. It's most profound.

I imagine Glennie nodding her head vigorously. She has long-maintained we hear with the whole body; the whole body can hear.

René Magritte's *The White Race* (1937) is a painting of a bathing nude, brassy and bold in composition, where each body part is a definitive shape. The thighs are like bell pumpkins nesting in the sand, the belly a perfectly round birthing-circle, the smaller circles above it also exact, these ones breasts. On top of those breasts it gets really interesting. Two noses act as columns holding up a pair of lips. Making a maybe-neck of the bathing nude, the noses with the red lips also make a retro table, straight out of 1937. Upon that table, atop of those lips, is an ear, centralised as a nose would be on a non-misshapen face. The ear balances a single eye, the eye the apex of the bathing nude. I wonder about ear-as-nose. I wonder about ear-as-centre-of-face. Of course I am concentrating on the ear, and now I am breathing in through my nose and imagining it's an ear. Suddenly I want my left ear to be the centre of my face. I want to hear with my whole face; I want my whole face to hear.

In *The Sound of Metal*, Ruben Stone is losing his hearing. He plays drums in a heavy metal duo with his partner, Lou, so there's a lot at stake. (If you're losing your hearing – if you're losing anything – there is always a lot at stake.)

Ruben's a heroin addict, clean for four years, so going deaf could destroy his sobriety. Knowing this, he goes to a house for Deaf addicts, where he's first told to wake up early, grab a coffee and a doughnut and go into the house leader's empty study and sit. If sitting bores him, he can write. When Ruben walks into the study, the pen and paper on the otherwise bare wooden table is so confronting he smashes the doughnut, screams, pounds, laughs in a way that isn't funny. He calls himself a fucken idiot. He can't accept what the house leader is trying to gift him: moments of quiet.

I can no longer 'just sit'. I get lost in my head, in my ear. It's mostly fear and *all that noise.* You see, I haven't lost my hearing, I've lost my silence, and losing my silence means I've gained noise and that noise is making me lose my hearing, so I lie.

I have a special relationship with my ear that's multifaceted. It's melancholy and desperate, which is why I'm deeply attached to Van Gogh's *Self-Portrait with Bandaged Ear*, a painting he made a week after he left hospital for treatment of the psychotic episode that had forced a razor into his hand and urged him to slice off his ear. When I look at the painting, I see a man who's mighty low, cold in his own home, a fur-lined winter cap and heavy coat buttoned at the collar. The bandage protecting the area where his left ear had been a constant reminder of just how fragile he is in this world, a visual that would trigger sadness and dread, which he must sit with.

Diemut Strebe photographed Lieuwe van Gogh after his DNA procedure for *Sugababe* and also called it *Self-Portrait with Bandaged Ear*. Lieuwe has ginger hair and piercing blue eyes and clearly resembles his great-great-great uncle. Around the left side of his face is a modern-day white cup-and-strap bandage protecting his ear. He looks away from his would-be viewers, as does Vincent in his own self-portrait, and it's as if the young Van Gogh is being haunted by an old trauma, perhaps one that happened more than a century ago. I see Lieuwe. I see Vincent. I see my ear.

✱ *A whole body mystery*, p. **113**
　The psychedelic remedy for chronic pain, p. **195**

A UNIVERSE SEEN BY WEBB

Sara Webb

If you ask an astronomer, 'Are time machines real?' you might be surprised to hear the response: 'Of course!' Now, I'm sadly not talking about a big blue phone box (Dr you-know-who) or speedy Delorean (Back to the you-know-when). Instead, I'm talking about telescopes. And the James Webb Space Telescope (JWST) is the most amazing time machine we've built yet.

Orbiting around the Sun at a distance of about 1.5 million kilometres from Earth, the JWST is seeking to answer some of the biggest questions in science. What were the first stars? How do galaxies evolve? How do stars evolve and could there even be life out on distant exoplanets?

To do this, JWST uses the power of science to look back over billions of years of universal history.

In this wonderful universe of ours, we have several laws of physics that govern throughout. One of these, and my personal favourite, is the speed of light. A universe can have a speed limit, and in our case, it's how fast light can travel in a vacuum. Light, no matter what its energy is, can only travel at the peak speed of 299 792 458 metres per second. Mighty fast from our perspective, but pretty slow in the grand scheme of the universe.

Every time we look up at the night sky, we are looking back in time, to space as it once was. Even our own Sun appears as it was a whole eight minutes earlier. For our brightest stars, this time gap might only be a few hundreds or thousands of years back in time. But for distant galaxies, it can be billions and billions of years. This is what the JWST was built to do: to look back into time, like never before, and unveil the beginnings of our universe.

As we see it from Earth with our own eyes, the night sky is only the tip of the iceberg. Our galaxy and the galaxies beyond it often need much better 'eyes' than our own – these come in the form of specialised telescopes, which can see different types of light than human eyes. To really get 20/20 vision we need to leave the fuzz of our atmosphere behind. JWST does just this. It allows us to peel back how galaxies looked and evolved, right back to the beginning.

To see light from very distant galaxies, we need a very large collecting area. As light travels through space, it follows yet another law: it decreases in brightness over the amount of space it has spread into. We can visualise how this works when we use a flashlight – the beam spreads out but appears to look fainter the further away it is. This is exactly what is happening on universal scales. The incredible mirror of JWST is designed to try to capture as much of that distant light as possible.

JWST's beautifully complex mirror measures a massive 6.5 metres across and is composed of 18 gold hexagonal-shaped segments. They're not just gold in appearance: they're actually covered in a microscopically thin layer of gold. This layer is key for optimising how much infrared light can be reflected into the camera. The JWST mirror can reflect approximately 99 per cent of the infrared light that hits it, meaning it has excellent infrared eyes in space. JWST is about 100 times more powerful than the mighty Hubble Space Telescope. Pretty impressive!

Are you wondering why infrared light? There are a few reasons. First, the amount of infrared light we can see here on the Earth is very limited. Our atmosphere emits infrared light and is effectively opaque to that coming from space. Second, and the most important reason, is that in the very distant universe infrared light is all that's visible, all thanks to something called cosmological redshifting.

For almost 100 years now, we've appreciated that the universe was so much bigger than our 'home' Milky Way Galaxy. In the

1920s Edwin Hubble worked tirelessly to estimate the distances to our extragalactic neighbours, but he made a puzzling discovery along the way.

He found that not only were galaxies moving away from us, but the further away the galaxy was, the faster we were moving from it ... meaning the universe was expanding. This is the reason we need JWST to 'see' in the infrared.

What's causing the universe to expand is dark energy – a cosmic conundrum to cover another time. For now, we just need to know that the universe is constantly getting bigger, and that this expansion of space does a pretty awesome thing to light: it stretches it!

Light which began blue will turn red over time. This is due to energy loss and it happens across the entire electromagnetic spectrum. In the local universe, the redshifting of galaxies is enough to detect only small shifts towards the redder wavelengths. In the distant universe, the effects become apparent. Whole galaxies seem to 'disappear' in visible and UV wavelengths at a certain distance. That light is still there, but now mainly in the infrared. This is where JWST can shine, as it can capture galaxies that Hubble never physically could.

And wow. JWST has already outdone all our expectations.

In the first images released, the thing astronomers were most excited about wasn't the beautiful big galaxies – it was the almost-invisible little tiny dots in the background. To astronomers, these small points of light are like the best gift the universe could give us. They are the most distant galaxies in our observable universe.

Just from those first few images, we already have a handful of candidate galaxies ready to take the crown for the most distant galaxy. If these galaxies are confirmed, their light is well over 13.5 billion years old. Emitted only a few hundred million years after the beginning of the universe, the light in these galaxies is most likely from the first-ever stars, ones made entirely of hydrogen and helium.

What's truly mind-blowing about these first galaxies is how far away from us they now are. The light has taken 13.5 billion years to reach this side of the universe, but those galaxies are much further away than that. Due to dark energy accelerating the expansion of the universe, these distant galaxies now sit 35+ billion light years away.

JWST allows us to travel that expanse of space and time. It allows us to 'see' regions of galaxies that were once 'invisible' to us. Infrared light is an exceptional probe of dust in galaxies. This is because galactic dust is very good at absorbing and scattering blue light and emitting redder light.

The best example of this is the breathtaking image of the unassuming galaxy NGC 628. Previous Hubble images of this galaxy had people in awe of its perfect spiral structure. JWST's infrared images of this galaxy left us speechless. It looks like a portal to another universe, or a gate to another dimension. The spiral structure was so much more detailed than we could ever imagine. By probing the light from dust and gas, JWST can give a different perspective on the galaxies we once thought we knew well.

It is even providing us better insight into our very own galaxy, revealing to us areas that we've only previously dreamt of seeing. The Carina Nebula was captured with all the power of JWST and it's nothing short of absolutely spectacular. It's my favourite space image, ever. This image of the cosmic cliffs uncovers hundreds of previously unknown stars, those newly formed in this enormous stellar nursery. The amount of dust and gas in nebulas like these makes them effectively opaque to optical telescopes. Being able to peer behind the dust in complex areas like this in the Milky Way will help us better understand the birth and evolution of stars.

This is only the very beginning of the incredible science to come from JWST. The minimum mission time will be five years, but NASA has an exceptional track record of missions outdoing their expected lifetimes. Hubble, designed for a 10–15-year mission, has been in orbit since 1990, and is still going strong 32 years later.

There are four main science themes of the JWST mission: 'Early Universe', 'Galaxies Over Time', 'Star Life Cycles', and 'Other Worlds'. Within these four umbrella science themes, JWST will help us uncover the secrets of the universe. These questions will be investigated around the world, with thousands of scientists involved in 14 countries.

Here in Australia, we have our very own superstar astronomers already working hard on this data. Professor Karl Glazebrook is leading a team out of Swinburne University of Technology to use machine learning and artificial intelligence to quickly work through the mountains of data to come, looking at galaxies and how they formed in the very early universe. Dr Benjamin Pope from the University of Queensland is working on an entirely different science goal, studying dead stars, planets and even asteroids, using new and exciting techniques.

We are already seeing an incredible amount of JWST-prompted science appearing in pre-print on the ArXiv, and the competition for most distant galaxy is nowhere near over. And we can look forward to seeing results soon from the many other amazing projects currently ongoing.

What a time to be alive!

✱ *Galaxy in the desert*, p. **49**
  First Light, p. **83**
  Antimatter: how the world's most expensive – and explosive – substance is made, p. **297**

FIRST LIGHT

Meredi Ortega

Launch of the James Webb Space Telescope, 25th December 2021

The glam rock music vid of it. TV dish lashed
to silver-mauve Kapton kite a million miles away of it.

This gift is very nominal, nominal and gold.
Did you put it in an airless, frigid vault? Did you blast

and bone rattle it, try out the 'Try Me'? Keep
opening and closing microshutters until with eyes

shut you could see an infinity of tiny windows.
It's enough to keep an astrophysicist awake at night.

Pirates and the three hundred and forty-four things
that might or might not happen. Did you unbox

it in the cleanroom? Did you move spotlessly
about? Aligning and folding and pinning the shield

with wire ties like an ice moon fashion doll that
will glide from its casing in a gown of white rocket.

The thick eyeliner of its frill. Starlight reveller, time
traveller, otherworld teller, eclipser of suns.

Only to think of it circling in space, its great back
always turned on us like we are unloved.

This gift is so hexagonal, hexagonal and bold.
The sheer cold mirror of it. Rare beryllium from Utah

honeycombed and polished to perfection of it.
Collapsed like a buggy with the baby still inside, safely

stowed in the nose cone in a jungle of lightning towers.
In a jungle. In coldest clasp at cryo-arm's length

suckling fuel until the last. All systems are green, back-
wards counts Jean-Luc Voyer. 'Décollage!'

✻ *Galaxy in the desert*, p. **49**
   A universe seen by Webb, p. **78**

A SUBANTARCTIC SENTINEL

Drew Rooke

In a back corner of the Royal Tasmanian Botanical Gardens in Hobart is an unassuming but very special teardrop-shaped pale yellow building. Made of thick concrete and roughly 14 metres long by 6 metres wide, it is reminiscent of a small wartime bomb shelter. Grey rocks and grasses line its base and above its double-layer clear polycarbonate roof is a white metal frame that supports a large sunshade. Beside the glass entrance door is a sign inviting visitors to 'come inside and step ashore on a subantarctic island'.

The subantarctic region, located a few degrees of latitude either side of the Antarctic Convergence, mightn't have the same mythical status – the same hold on the collective human imagination – as the icy continent further south does, but it does contain numerous islands that are havens of life in the middle of the huge oceanic desert that is the Southern Ocean. The Subantarctic Plant House, as this building is known, is based on one of these islands: Macquarie Island, a cold, wet and windy speck of land located halfway between Tasmania and Antarctica.

The carefully controlled environment inside the Subantarctic Plant House attempts to re-create the one on Macquarie Island – or 'Macca', as it is more affectionately known. It is grey and dim. Freezing air blasts out of four vents beside the door and occasionally a cool mist blows around the room. Covering the wall is a hand-painted mural featuring scenes from the island: petrels nesting on cliff faces, king penguins waddling across a shingle beach pounded by huge waves, and rain pouring down onto a cloud-shrouded plateau. The stream that trickles through a patch of mire into a small pool and the field recordings of grunting and bellowing elephant seals playing from speakers make it even easier to imagine that you have indeed suddenly stepped ashore on Macca.

As does the collection of beautiful plants – the lush green ferns, megaherbs, mosses, grasses and liverworts. Situated on either side of a small wooden boardwalk that leads around the room, they – along with the rocks and gravel they are growing in, over and between – were all collected from Macquarie Island in the 1990s. Having evolved in such a harsh environment, none of them are particularly large; among the biggest is *Stilbocarpa polaris*, a megaherb commonly known as Macquarie Island cabbage that reaches only a metre in height and whose saucer-sized, leathery leaves are covered in fine white hairs – just like the thin stems they grow on – that help trap heat.

But the Subantarctic Plant House isn't a complete microcosm of the flora on Macquarie Island. One plant that is missing is the endemic *Azorella macquariensis*. Once confused with the closely related *Azorella selago* found on other subantarctic islands, it is a very slow-growing and very compact semideciduous cushion plant with diminutive flowers that lives in just the top 2 centimetres of soil and depends on a consistently wet and misty environment to thrive. In many parts of Macquarie Island, it forms giant, iridescent green terraces and undulating carpets that can be hundreds of years old and collect significant amounts of carbon, prevent erosion and support a diverse range of epiphytes and microarthropods. Because of this, the plant is considered a keystone species – an ecosystem engineer – just as eucalypts are in so many forests throughout the Australian continent.

There was a brief period when the Royal Tasmanian Botanical Gardens did have one specimen of *A. macquariensis* in its sub-antarctic collection. But – as Lorraine Perrins, curator of the Conservation Collections and Subantarctic Flora at the gardens explains – mimicking the exact environmental conditions found on

Macquarie Island is 'practically impossible' and as a result the plant 'struggled'; it was very etiolated and did not form the tight canopy that characterises the species. Then, in 2011, the cooling system of the Subantarctic Plant House broke down and was disabled for several weeks. Despite their best efforts, Perrins and her colleagues could not save the already struggling *A. macquariensis* specimen. Its death, Perrins says, was 'devastating'.

It was also symbolic of a far more devastating and ongoing rapid mass mortality event affecting the entire population of *A. macquariensis* on Macquarie Island. Although much mystery still surrounds the widespread dieback of the cushion plant and the associated ecosystem collapse, scientists know enough about it already to confidently say it is further proof that nowhere, no matter how protected or free of other environmental pressures, is safe from catastrophic climate change – and to fear for the future of other subantarctic islands and the rare and beautiful life forms they are home to.

The dieback of *A. macquariensis* is another tragic chapter in Macquarie Island's extraordinary history.

Roughly 30 million years ago and 4000 metres below sea level, the Indo-Australian and Pacific tectonic plates started to tear apart. Molten rock surged through the rift between them and solidified as it cooled, forming a new oceanic crust. Over time the two plates started to crash together, creating a ridge in the crust that gradually rose through the lightless depths until, approximately 600 000 years ago, its crest broke through the waves.

Since that moment, the piece of raised oceanic crust that is Macquarie Island has continued to be uplifted at a rate of several millimetres per year; characterised by an undulating alpine plateau dotted with lakes that sits atop steep scarps, its highest point is presently 435 metres above sea level. Although the island is only 34 kilometres long and 5.5 kilometres wide, the vast submarine mountain range of which it's the subaerial tip stretches 1600 kilometres from the South Island of New Zealand to near the Antarctic continent. Known as the Macquarie Ridge, it is significantly steeper than the Himalayas, the Andes and all of the world's other great terrestrial mountain ranges.

Being the only known place where rocks from the Earth's mantle, 6 kilometres below the ocean floor, are being actively exposed above sea level – and that contains a complete and distinct sequence of rocks from the upper mantle to the upper crust – Macquarie Island is invaluable to understanding the geological features and processes of oceanic crust formation and plate boundary dynamics. Indeed, its unique geomorphology was one of the main reasons why, in 1997, UNESCO agreed that it had outstanding universal value and inscribed it on the World Heritage list.

Another was the array of flora and fauna it supports. Despite its small size, Macquarie Island is home to nearly 50 vascular plant species as well as 86 species of mosses and 51 species of liverworts – all of which arrived via long-distance dispersal on the wind or on the wings of birds. It also provides critical feeding and breeding grounds for a plethora of wildlife, including 100 000 seals and four species of penguins totalling four million birds that form vast colonies on its shingle beaches and tussocky coastal platform and slopes.

Since the ageing research station on the isthmus was built in 1948, Macquarie Island has been used as a base for globally significant scientific research. But earlier human visitors to the island used it for a very different reason.

One of the earliest human visitors was Frederick Hasselborough, who was captaining the *Perseverance* on a voyage to sealing grounds south of New Zealand in July 1810 when it was blown off course. While the brig was lost, Hasselborough spotted an island in the distance. Although evidence was later found indicating Polynesian sailors had previously visited this island, Hasselborough believed it was undiscovered and took the liberty of naming it in honour of the then governor of New South Wales, Lachlan Macquarie.

Upon seeing that the island's shores were teeming with fur seals, Hasselborough put a small gang of men ashore to begin harvesting their pelts while he sailed north to Sydney for extra provisions, determined to keep secret what he had found. He succeeded in doing so until – as the tale goes – the evening before he returned to Macquarie Island, when he attended a farewell party that had been organised by a merchant and sealing master named Joseph Underwood.

Unbeknown to Hasselborough, the party was part of Underwood's plan to try to dupe him into revealing the location of his recent discovery. When it was in full swing, Underwood proclaimed that he had known about the island for several years and bet Hasselborough £20 that he could name its precise position. Too drunk on rum and his own ego to realise he was being hoodwinked, Hasselborough accepted the challenge and doubled the stakes before writing in chalk on the underside of the table – as had been agreed – the island's exact coordinates: 55°S, 159½°E. Underwood then took the chalk and, pretending to stoop down to write his entry in the competition, simply looked at the figures Hasselborough had written. He feigned disappointment, crying out 'I have lost', but he knew the knowledge he had just gained was worth significantly more than the £20 he gladly handed over to Hasselborough.

It didn't take long for this knowledge to spread wider – and for Macquarie Island to become an open-air slaughterhouse. By December 1810, another three Sydney-based sealing gangs were operating there and, within the first 18 months of operation, roughly 120 000 fur seals had been killed for their fine pelts. Within a decade, fur seals were effectively wiped out on the island and sealing gangs then turned their clubs on the elephant seals and, later, penguins whose blubber was stripped and boiled down for the oil it contained, which was shipped around the world to illuminate houses and entire cities.

In 1920, following a sustained environmental campaign led by Sir Douglas Mawson and other members of the Australian Antarctic Expedition who established the first major scientific base on Macquarie Island in 1911, the commercial exploitation of its wildlife finally ceased. Thirteen years later, it was formally declared a wildlife sanctuary, and the seal and penguin populations slowly started to recover. But the sealers left behind not just their huge rusting try-pots, digesters and boilers but also a destructive environmental legacy in the form of numerous invasive species, such as mice, rats, cats, wekas and rabbits, which voraciously preyed on seabird chicks and eggs and stripped entire coastal slopes of native vegetation. This led to significant erosion and loss of nesting habitat long after the last penguins and seals had been clubbed to death.

Efforts to eradicate the huge number of invasive animals on Macquarie Island began in earnest in the mid-20th century. Cats were finally eradicated from the island in 2000 following a 15-year program (although not before they had helped drive two native birds – the Macquarie Island parakeet and the Macquarie Island rail – to extinction). But this had the unintended consequence of causing rabbits to dramatically multiply, as they were no longer preyed upon, prompting the establishment in 2007 of a new $25 million project to simultaneously eradicate rabbits, mice and rats from the island using a combination of baits, skilled hunters and specially trained dogs. Seven years later, the project was declared a success, making it the largest successful island pest-eradication program ever attempted.

This was a major cause for celebration among the scientists working on the island. But by then they had a fresh cause for major concern: the widespread dieback of *Azorella macquariensis*.

As when Frederick Hasselborough stumbled upon Macquarie

Island, Dr Dana Bergstrom discovered the die-back by chance. An award-winning applied ecologist with the Australian Antarctic Division, Bergstrom has soft blue eyes, curly salt-and-pepper hair, and a vast knowledge of the Antarctic and subantarctic regions that she's gained over the nearly three decades she's spent researching them. She first visited Macquarie Island – 'a little jewel in the ocean' – as a 21-year-old master's student in 1983 to research fossil peats. The trip was challenging – she didn't have proper boots or a proper raincoat and suffered hypothermia – but it left a deep and lasting impression on her.

'I fell in love with the place then,' she says. 'It got into my blood.'

On many subsequent research trips to Macquarie Island, Bergstrom became very familiar with *A. macquariensis*. To her, it is one of the most bizarre plants on the planet – 'like a whole bunch of balloons blown up with water and sandwiched together' – and one of the most beautiful. She likens the vast terraces and carpets it forms over the landscape to 'old-growth forests'.

One specimen of *A. macquariensis* Bergstrom became particularly familiar with was located on the northern end of the island, near the beginning of the descent from the plateau to the research station. Shaped like the island itself, it always gave her joy whenever she spotted it on her way back from being out in the field. 'No matter how bad the weather was, you'd see it and think, "I'm almost there." It was a little signpost, which is helpful because sometimes you can't see ten metres because of the fog and the mist.'

In December 2008, Bergstrom was on the island mapping vegetation when she saw that this particular plant was not vibrant green – as it should have been at that time of year, having emerged from its winter senescence – but brown and covered in holes. From this she knew the plant was dead. And once she had investigated further, she noticed there were many other cushion plants that had suffered the same fate. So too had the mosses and megaherbs that depend on *A. macquariensis* for their own survival.

'It was a really *holy fuck* moment,' Bergstrom says. 'No one had seen it before because it is such a patchy landscape. It's only because I had been going there since I was twenty-one that I was able to notice what was going on.'

Back at the station, Bergstrom picked up the phone. The person she called was Dr Jennie Whinam.

Like Bergstrom, Whinam's connection to subantarctic islands stretches back a long time. Her first visit to one was in 1987, when she travelled to Heard Island, 4000 kilometres south-west of Perth, as part of her PhD researching the ecology of peats. This was, she says, 'the start of a lifelong passion for subantarctic islands'. The following year, she visited Macquarie Island for the first time and over the next three decades travelled to subantarctic islands roughly once every two years as part of her work with Tasmania's Department of Primary Industries, Parks, Water and Environment.

'Subantarctic islands are the places that grab my heart,' she says. To her, they are 'wild and sublime'. Indeed, she loves them so much that when she visits Hobart from her home, a 30-minute drive away, she will often make a concerted effort to go to the Subantarctic Plant House at the Royal Tasmanian Botanical Gardens, 'just to have that feeling of instantly being back'.

Whinam thinks of the vegetation on Macquarie Island as being like a 'more extreme version of a Japanese garden. Everything is in miniature, so beautifully in place.' And for her, the vast rolling cushions and terraces of *A. macquariensis* are intrinsic elements of the island's floristic beauty. So when she answered Bergstrom's call that day in December 2008 and learnt about the dieback, she was 'immediately' alarmed – especially because she had been on the island the previous year and hadn't observed a hint of it at any of the sites Bergstrom had.

By the end of the following year, Whinam's alarm had grown significantly after further surveys conducted by her and her colleagues revealed dieback at 88 per cent of 115 *A. macquariensis* sites. The worst examples of the dieback were located in the northern section of the island, where many of the cushion plants had completely died and been blown or washed away by the wind and rain, leaving nothing but gravel, or had been colonised by native grasses such as *Agrostis magellanica*.

This data ultimately led to the Australian Government listing the cushion plant as critically endangered in August 2010. 'As far as I know, it was the first plant species to go from non-listed to critically endangered in one leap,' Whinam says.

The rapid speed at which the dieback was occurring was particularly shocking given that the incredibly harsh environment of Macquarie Island generally means that, as Whinam explains, 'things grow slowly, and they die slowly. Until you get to a tipping point, and then everything suddenly falls over'.

Bergstrom and her colleagues hypothesised that the primary cause of this particular tipping point for *A. macquariensis* was 17 consecutive years of drought that occurred in the larger context of Macquarie Island's changing climate: over the past four decades, it has seen increased but more episodic rainfall, higher wind speed, less cloud cover and more sunshine hours, which has made it drier and warmer overall. This is problematic for a species such as *A. macquariensis*, which is particularly vulnerable to climate change because, as Bergstrom explains, it 'has just thrown away features that would make it able to survive in drought'. Weakened and water-stressed, the cushion plants were then more susceptible to pathogenic infection, which the chlorosis band on their canopies suggested was occurring.

In 2009, Whinam conducted sampling on multiple sites across Macquarie Island to try to identify the particular pathogen – or pathogens – in question. 'We were under a lot of pressure because people wanted answers,' she recalls. It didn't help that poor weather limited the amount of time she and colleagues were able to spend in the field collecting samples. 'We were frantically digging in the gravel and mud while it was pouring rain and howling wind,' she recalls. 'It was no spring picnic.'

A wide range of pathogens were detected in the samples Whinam collected. One that popped up in multiple samples was *Pythium splendens*. But like all of its counterparts, it wasn't known to be fatal on its own. Adding to the mystery was that no evidence was found indicating that the rangers working on the pest eradication program had introduced a foreign pathogen to the island and were spreading it as they scoured the landscape for rabbits, rats and mice, which was what Whinam, Bergstrom and their colleagues had initially feared.

Over the next few years it became clear that the mysterious soil-borne pathogen had overtaken drought as the predominant cause of the dieback, with the telltale 'yellow line of death', as Whinam calls it, sweeping across the compact green canopies of more and more cushion plants. Although they didn't know what the pathogen was, scientists such as Whinam came to believe that it had been latent on the island until it had started to flourish as a result of the recent shift in climate.

As concern about what was happening to *A. macquariensis* mounted, so too did the quest to properly quantify the health of the entire population and work out a conservation plan for it. Indeed, this was the focus of a multiyear collaborative research project launched in 2015. Funded by the Australian Antarctic Division, the project was managed by Professor Melodie McGeoch from the Department of Ecology, Environment and Evolution at La Trobe University. Whinam and Bergstrom were among those involved in it, as was Dr Catherine Dickson, an experienced applied ecologist with a passion for threatened species and a background in environmental conservation in South Australia, who came on board as the lead field researcher.

Dickson's first field trip to Macquarie Island as part of the project was in the summer of 2016–17 to survey the extent of the dieback across the island and determine the distribution and abundance of potential refugia for *A. macquariensis*. It was an especially wet and windy trip, with three monthly weather records on Macquarie Island broken in December 2016, and Dickson quickly learnt that conducting field work on a subantarctic island comes with unique risks – like falling into a putrid elephant-seal wallow full of mud, urine, faeces, shed hair and moulted skin.

By the end of the trip in February 2017, Dickson had, with the help of colleagues including Bergstrom and Whinam, deployed microclimate data loggers, collected leaf samples and surveyed the cover and dieback of *A. macquariensis* at 62 sites in various locations across the island – from very steep exposed slopes to more gentle sheltered ones and to fellfield peaks. The following summer she returned for two months to collect the microclimate data loggers and resurvey the condition of the critically endangered cushion plant at an additional 20 sites.

By then, much of the island's vegetation had recovered significantly, since rabbits had been eradicated nearly a decade earlier. But Dickson's work – for which she was ultimately awarded her PhD – revealed a more sobering story for *A. macquariensis*. She found, for instance, that the dieback of cushion plants was ongoing but highly variable across the island, with some surveyed sites showing near-complete death of cushions and associated mosses, liverworts and megaherbs; that it had moved southward and was now centred on the middle of the island; and that it was more prominent in areas with a low number of freezing days and very high humidity – conditions known to be conducive to pathogenic infection and that are expected to become more common on Macquarie Island under climate change.

The research project officially came to an end in October 2019. Since then, no further major research into the dieback of *A. macquariensis* and the associated ecosystem collapse has been conducted, and the exact pathogen now believed to be driving the dieback remains unknown.

On top of Wireless Hill, located at the far northern end of Macquarie Island and so named because it was the site of a wireless radio station that linked the Australasian Antarctic Expedition's main base in Antarctica with Hobart between 1911–14, there is an *ex situ* population of 54 *A. macquariensis* plants growing in large white, irrigated buckets. Collected from different sites around the island in 2010 and 2013, these plants are carefully monitored via monthly photographs. At present they are in good health and are considered one of the best ways of conserving the species.

Another is the *A. macquariensis* seed bank stored at the Tasmanian Seed Conservation Centre, located in the Tasmanian Royal Botanical Gardens, a short stroll away from the Subantarctic Plant House. The seed bank contains just over 4000 seeds, which horticultural botanist Natalie Tapson meticulously harvested by hand from 365 healthy cushion plants on Macquarie Island over several weeks in 2016. However, there is one major problem with the seeds: to date, scientists have been unable to work out how to break their inbuilt dormancy – to replicate the precise combination of conditions required for them to germinate.

But even if scientists are able to crack the secret germination code someday in the future – and even if they are able to identify the mysterious lethal pathogen and can then develop an effective fungicide – it is very unlikely that *A. macquariensis* will ever return to its former glory on Macquarie Island. That, at least, is Bergstrom's belief, and she bases it on the fact that no serious action is being taken to mitigate the underlying driver of the dieback and the colonisation of areas once dominated by *A. macquariensis* by more resilient native grasses – that driver being climate change. Indeed, since the dieback was first discovered more than a decade ago, Macquarie Island has gotten even hotter: on 8 February 2022, it recorded its highest ever temperature of 17 degrees Celsius, more than 8 degrees hotter than the mean temperature for that time of year. The previous maximum temperature recorded on the island was 14.4 degrees Celsius in December 1984.

'We're so used to thinking we're in control of nature and can come up with solutions to problems,' Bergstrom says. 'But I don't think there is a solution here. I don't think there is a happy ending.'

Looking into her scientific crystal ball, the future she pictures for *A. macquariensis* – a species whose genetic signature suggests it has grown on Macquarie Island for roughly 200 000 years – is one where the ancient cushion plant will be locally extinct in many areas, confined to the few pockets with colder and wetter microclimates than elsewhere. Where it once formed vast terraces and carpets, there will either just be grassland or bare ground – a change that is already clearly visible across the island.

According to Bergstrom, this represents not only the transformation of an entire ecosystem but also the loss of 'a defining component of the intrinsic wild value of the island'. It is a loss that she speaks about with tears in her eyes and a faint tremble in her soft voice – clear signs of 'solastalgia', the term coined by environmental philosopher Glenn Albrecht to refer to the 'form of psychic or existential distresses caused by environmental change'. And it is one that may very well be the first of many irreversible losses that occur throughout the subantarctic. As Whinam says: 'It's very possible that what we're seeing on Macquarie Island, we might soon start to see on Heard Island or the other incredibly rich, biodiverse specks in the Southern Ocean.'

That this knowledge doesn't seem to have any material impact – that so few people seem to care about it – can fill those who have been responsible for generating it over many decades with despair. 'Sometimes I throw my hands up and say, "Fuck it. I'm just going to grow plants and spend all of my time in my garden",' Bergstrom says.

But – at least for the moment – she isn't going to do this: 'Because that's being entirely self-focused.'

Instead, Bergstrom and so many of her colleagues continue to direct their focus elsewhere: to science, to improving our understanding of how subantarctic and Antarctic environments are rapidly changing in a warming world – and, by extension, to holding us all to account for the environmental consequences our collective actions are having, including on one of the wildest and remotest pieces of land on Earth.

✱ *Buried treasure*, p. **7**
A city of islands, p. **60**
Onboard the space station at the end of the world, p. **283**

MODEL OR MONSTER?

Amalyah Hart

In a two-room laboratory sequestered in a hunkered-down building in Werribee, Victoria, a small but mighty group of baby frogs, some of the last bastions of their embattled species, are patiently waiting to die.

I've come to visit them on a blinding hot February day, and I've been excited about the encounter for weeks. But when I open the door and see them, squatting blithely in row upon row of plastic tanks, I'm struck with a potent wave of tragedy.

These are juvenile southern corroboree frogs, tiny little things with an overlaid pattern of bright yellow and black on their breathable skin, like a croaking nuclear waste sign. That's pertinent, because they secrete a poisonous alkaloid through that skin that can kill prospective predators.

I squat on my haunches, greeting one of the little frogs as it hovers, suction-cup toe-pads pressed against the glass, its little throat moving rhythmically up and down. It has no idea what's coming.

These young frogs, some no bigger than the tip of my thumb, are doomed – they won't live for more than a few months at most. But their deaths may be the key to reversing the march of their species towards extinction.

That's because these frogs are the first test subjects in a project that will plumb some of the most exciting (and controversial) realms of science, in the quest to conserve a dying species – by striding into the vanguard of gene editing.

Lee Berger was in the midst of her PhD at James Cook University in the 1990s when she began investigating an alarming and inexplicable global decline in amphibian populations that had been going on for at least 20 years. At the time, its effects were seen most acutely in the rainforests of Central America and Queensland.

Mass frog deaths were moving across the landscape, following the kinds of patterns you expect from an epidemic, and researchers were proposing that some sort of exotic pathogen must be behind the deaths. But it was Berger who, in 1998, first identified a fungus, *Batrachochytrium dendrobatidis*, suffusing the skin of sick frogs.

Southern corroboree frogs are native to the mossy sphagnum bogs of the northern Snowy Mountains, and are exclusively found within the limits of Kosciuszko National Park. They summer in mossy chambers, where the males voice a charmingly offbeat 'squelch' to attract their mates. In the winter, they retreat to the snow-sheltered undersides of snow-gum logs and leaf litter.

Southern corroborees are now functionally extinct in the wild, because *B. dendrobatidis* causes a ravaging disease known as chytridiomycosis. According to Zoos Victoria, there may be fewer than 50 of these frogs left in their natural habitat, and the tiny group remaining only persists because breeding programs have periodically replenished their dwindling population.

B. dendrobatidis, also known as the chytrid fungus, takes advantage of amphibians' most important evolutionary quirk – their porous skin. 'Frogs actually absorb oxygen through their skin, and they also absorb a lot of electrolytes that way,' explains Tiffany Kosch, a research fellow at the University of Melbourne and one of the lead architects behind a daring new plan to save the frogs.

'So, how the fungus ends up killing them is they actually have a heart attack, because they're not able to maintain the correct electrolyte balance in their bloodstream, so the heart slows and they eventually die.'

And the chytrid fungus is a serial killer: by some estimates, it may be responsible for the greatest disease-driven loss of biodiversity in recorded history. It travels in water, spreading through interlocking stream systems and from frog to frog, decimating the populations it comes into contact with.

In the Melbourne research lab, I'm introduced to chytrid by the scientist who first illuminated its crimes. More than 20 years on from her initial discovery, Berger is now working in the same facility as Kosch where, in a small fridge in the next-door room, she carefully stewards petri dishes of the murderous fungus.

She takes one out, places it under a microscope and invites me to look.

Tiny little swimmers dart around on invisible flagellae, dodging bigger, globule-like cells that hover, suspended in the artificial light.

The swimmers are the infective cells, which latch onto the frog's skin and begin to multiply, transmuting into the fat globules that clog the frog's life-giving pores. Looking at these tiny cells, floating inoffensively on the microscope's plate, it's hard to believe they could wreak such ecological havoc.

The chytrid fungus has resisted more than two decades of efforts to remove it, and the disease kills around 95 per cent of the southern corroborees it comes into contact with. Kosch and her team at Melbourne University are aiming to reverse this seemingly intractable situation, by armouring the frogs with a set of genes that might protect them.

This will require one of two approaches. The first option is artificial selection: breeding creatures with the necessary genes together to produce a resistant population over several generations. Humans have been practicing artificial selection for millennia, ever since the first domesticated crop or tamed wolf.

The second approach is a bit more controversial. Synthetic biology is an umbrella term for a number of techniques that manipulate genes or sets of genes to achieve a desired result.

These include transgenesis, which involves transferring whole genes (or sets of genes) into one species from another species, and gene-editing, which involves 'snipping' out certain genes (or parts of a gene) and replacing them with others.

Synthetic biology can alter an organism's genotype (its genetic material) to produce a desired phenotype (the observable traits coded for by the genotype).

'So many of these frogs are now bred in captivity, but whenever they release the frogs they usually don't survive, because the pathogen can't be eradicated,' says Kosch. 'So, our idea is to test synthetic biology methods, which are very successful in agriculture, and see if they might work for conservation.'

About 10 000 years ago, as the last ice age relinquished its grip and *Homo sapiens* stood on the threshold of the Neolithic, archaeologists think that a handful of communities living between the Tigris and Euphrates rivers in Mesopotamia (modern day Iran, Iraq and Syria) began to play around with the seeds of wild plants, purposefully selecting and planting them for a predictable crop, and the certainty of food. In doing so, they could not have known that their selective choices would alter the genomes of entire species.

The agricultural revolution completely reshaped the structure of human life, enabling more permanent settlements and larger communities. And the crop plants that would emerge became both genetically and phenotypically distinct from their wild counterparts.

Successive civilisations have been meddling with DNA for millennia. What makes synthetic biology so different?

Current gene editing techniques allow scientists to make the kind of changes in a single generation that might once have taken many iterations of environmental trial-and-error. Proponents argue they also allow for much more precise changes.

'The really good thing about gene editing, compared to artificial selection, is you're only changing the genetic elements that you're trying to change, not a whole bunch of other things,' explains Kosch.

'And that's really important, because we want these frogs to be able to survive in the wild in future. And because we don't know what these frogs are going to be experiencing in the future, we have to preserve their genetic diversity.'

But that also makes it an impressively powerful tool, in the hands of a species that has not always been judicious.

'Whenever you have a technology that's really, really powerful, you can use it to do really good things, but you might use it to make big mistakes, or some people might use it with bad intentions,' says Christopher Gyngell, a biomedical ethicist at the Murdoch Children's Research Institute who has worked extensively on the ethics of gene editing.

'That's the crux of this issue, we've got this power now, and we have to decide how to use it.'

For Gyngell, the important thing in this type of research is to be slow, cautious, and reasoned.

'I think you need to be careful with these technologies, you need to move slowly,' he says. 'What human history has shown us is that ecosystems are really complex, and sometimes humans don't understand them as well as we think we did.'

One of the core criticisms of using gene editing in conservation, according to Kosch, is the idea that it's a band-aid over the problem, and that scientists and conservationists should instead focus on eradicating the threat.

The problem is that many of the threats to biodiversity around the world are now rampant – a Pandora's box of dire consequences.

Limiting global warming now won't stop or reverse its consequences for decades, at best. It's the same with chytridiomycosis.

Scientists have tried for decades to come up with ways to eradicate the disease, including expensive and dangerous chemical treatments with knock-on effects for other local species, and all to no avail.

From Kosch's perspective, then, the way to give this frog a fighting chance is to delve into its genes.

Only a few weeks after my first visit to the lab, I receive an email from Kosch: the experiment is underway.

The researchers have placed the frogs in little takeaway sauce cups. After pipetting a liquid mixture of water and chytrid fungus over them, they're kept in the containers for six hours, to allow the fungus to fully take hold. Then, over the course of the next few weeks, the team will observe how each frog is faring.

Kosch, like any good biologist, loves frogs, and she admits it's hard work to do. But she is focused on the research she believes is the species' best hope of survival.

'Obviously we don't want to be causing frogs to die, but at the moment unfortunately a lot of these little guys are not going to survive if they're released into the wild,' she says. 'If we're doing something that can maybe increase their chances of survival someday, it's worth trying.'

Around 95 per cent of the frogs will die within a few weeks of infection. The first indication that something is wrong is a kind of limpid sluggishness, a reluctance to move. The scientists put a gloved hand into the tank: a healthy frog will hop away, but a frog sick with chytridiomycosis will stay put.

Another tell-tale sign is posture: a healthy frog will sit with its legs tucked underneath it, whereas an unhealthy frog will splay its legs and struggle to climb.

In the final test, the researchers will take the frogs out of their tanks and place them upside down in the palm of their hand. A healthy frog will right itself, but a sick frog will lie there, helplessly supine. After this righting reflex is tested and failed three times, the frog will be euthanised.

The hope is that 5 per cent of the frogs will survive infection and recover, and their DNA could hold the key. By looking closely at the genomes of the survivor frogs, the team hope to identify the genes that are affording them resistance to the pathogen.

'This might be hundreds of genes, maybe thousands, or it may just be a few,' notes Kosch. 'At this point, we really don't know.'

But once they do know, the discovery will allow the researchers to work out how to arm new generations with the right genes that could help them survive in the wild.

So, do they use the old method or the new? Artificial selection or synthetic biology?

If the genes that code for resistance in these frogs are easily inherited, or involve many genes, all of which have a small effect, they're likely to pursue selective breeding, because it may be easier to breed the genes into a new population than to snip and replace so many, low-impact genes.

If the genes have low heritability, or they involve fewer genes that each have a significant effect, Kosch says they'll likely opt for gene editing, using CRISPR-Cas9.

That's because, in this case, it would be more difficult to breed a resilient population, and also because gene editing is much quicker – artificial selection can take tens or even hundreds of generations for the positive genetic change to take hold.

Both methods have risks: selective breeding in a small population can reduce genetic diversity, or could breed out other genes needed to survive in the wild. Inserting genes that carry markers for chytrid resistance could add another unforeseen weakness.

And there's another problem: releasing genetically engineered frogs into the wild is uncharted territory, so the methods they use will have to adapt as the regulatory landscape around this strange new world starts to take shape.

The chestnut tree is an icon of Americana; you'll have heard the Christmas song about nuts roasting on an open fire. The American chestnut tree was also a cornerstone species for local ecosystems up and down the US east coast for millennia.

'This tree was a very abundant mass producer,' explains William Powell, director of the American Chestnut Research and Restoration (ACRR) project at the SUNY College of

Environmental Science and Forestry, New York, US.

'It produced a lot of nuts for wildlife [and humans] to eat, it produced really straight-grained, rot-resistant wood for people, and the leaves were used as medicine by Native Americans as well as early settlers.'

It's a tree that provokes emotion for many of the people who live in its native homelands.

'It's ingrained into our folklore,' Powell says.

But the chestnut has been all but lost to the east coast, thanks to another invasive fungal marauder – *Cryphonectria parasitica*, or chestnut blight. First observed on US soil in 1904 in the New York Zoological Gardens, the blight was introduced when people began importing the Asian chestnut tree into the country.

'They didn't know at that time that when you bring a tree over, you bring all its microbes over also,' Powell says. While the fungus was a mildly annoying skin condition for the Asian chestnut tree, it spelled disaster for its American counterpart.

The fungus latches onto the tree's bark and starts to gnaw a wound, called a canker, into the trunk or branch. It does this by producing a substance called oxalic acid, which kills off the tissue in front of it to produce a necrotising substance that the fungus can eat.

The canker grows, eventually girdling the trunk or limb and cutting off the flow of nutrients; everything above the canker dies. If a canker forms at the base of the tree, everything above ground perishes.

It's estimated that over the last century, four billion American chestnut trees have disappeared because of the fungus, with small pockets remaining in the Carolinas, West Virginia and in Pennsylvania.

Chestnut trees can survive at the roots, so there's still a few million stump sprouts left. Periodically these stumps grow sprouts, which then encounter the fungus and die back, so the viability of a flourishing chestnut tree along the spine of the east coast is low.

But the existence of these stump sprouts means there's a relative wealth of chestnut genetic diversity, so conservationists seeking to breed a new (and perhaps improved) population aren't hampered by the problem that an endangered species tends to have: low variability in the population genome.

That's where Powell's research group comes in. For several decades now, Powell and his colleagues at the ACRR have been using genetic engineering techniques to alter American chestnut tree cells in the lab, with the goal of creating a tree that's resilient to the fungus – and it seems to be working.

They're utilising transgenesis, with a nifty little gene borrowed from the wheat plant. The gene produces an enzyme, oxalate oxidase, that detoxifies oxalic acid, breaking it down into carbon dioxide and hydrogen peroxide, two compounds the plant can use.

'And so, the nice thing is it doesn't actually hurt the fungus at all,' says Powell – who is perhaps more charitable than me. 'All we're doing is taking the weapon away, so now it can live as a saprophyte [decomposer] on the tree, it can still cause a little bit of damage, but it's not the severe damage where you get the whole tree being killed.'

They've transferred the gene into the tree cells using another handy bacterial tool, agrobacterium-mediated transformation.

'Agrobacterium is a bacteria that naturally genetically engineers trees or plants in general, and has been doing that for millennia,' says Powell. 'So that's what we did with the oxalate oxidase – we made what's called a vector, put it into the agrobacterium, and then let the agrobacterium put it into the tree.'

Once the gene has been introduced, the researchers kill off the bacteria with antibiotics. This is all done while the tree is still in its cellular stage in the lab, so it's tightly controlled.

The team have been field-planting these trees since 2014. But the next step – actually releasing them into the wild and populating the whole coast with its beloved chestnut once more – could take years of further regulatory hurdling.

'Those [field-planted trees] are all planted under USDA permits,' explains Powell. 'And they come out and inspect our fields, we have certain rules to follow, and we can't let them pollinate right now.'

At the moment, the program is under review by three agencies: the US Department of Agriculture (USDA), the Environmental Protection Agency (EPA), and the Food and Drug Administration (FDA).

Reviews for transgenic plants in the past in the US have almost exclusively focused on agricultural plants that are harvested every year; the chestnut, on the other hand, is not only a wild-dwelling plant but a long-lived tree.

The concept of changing an organism's genes and then releasing it back into the wild is intrinsically Frankenstein-ish to some. Ever since scientists carefully unpicked the double-helix 70 years ago, DNA has been something of a godhead – an inviolable biological structure.

But that's not always how scientists see it.

'Part of the root of some scepticism is that there's something kind of integral and wholesome about a genome, and that that kind of defines what a being is,' says Andy Newhouse, an ecologist and assistant director of the project at SUNY.

'But as we've been learning more and more over the past couple of decades about where our genes come from, they get swapped around all the time.'

Some of this resistance may stem from a perspective that humans and other creatures are all separate entities, rather than members of an interconnected community of organisms.

'There are so many genes in people, in plants, in chestnut trees and in corn from other stuff; from viruses, from bacteria, from related plants and unrelated plants,' Newhouse points out. 'So, I wish I could help convey that the genome doesn't define what a being is.'

So, what do Powell and Newhouse make of fears about Franken-tree?

'If you actually look at this kind of agrobacterium transformation compared to older, traditional breeding methods like hybrid breeding, it turns out this agrobacterium transformation causes ten-fold or maybe even 100-fold fewer changes to the genome than all those older techniques,' says Powell.

'And so, in reality, especially for conservation where you're trying to preserve the identity of certain species, it's kind of nice to have less of those changes.'

The next time I visit the Werribee frog lab, there's a chill in the air. I slip on a pair of gumboots and a lab coat at the door so I don't unwittingly track the fungus out of the room and commit biological warfare against other frogs in the building.

Of the 58 frogs I met before, just four are left – their lonely plastic tanks dwarfed by the empty shelf space where their comrades once sat. But there's a glimmer of hope to this otherwise sad story, because something quite remarkable seems to have happened.

One of the frogs – tank 27 – has escaped the inevitable. This frog doesn't seem to have contracted the disease at all, despite sitting in a liquid mixture of live chytrid fungus for six hours.

Given that Kosch only set out to identify frogs that could survive infection, the prospect of a frog that could actually be resistant to the infection itself is tantalising.

'We don't know if it's an anomaly or a super resistant frog, but it's a weird little thing,' she says.

It's possible that somehow this particular frog was given a quantity of dead fungus – but Kosch says that's unlikely. The next phase of the study, then, will show them whether any other frogs have this mysterious super-resistance.

Apart from super-frog and his three dying friends, the other 54 all succumbed to the bug, and were euthanised when they failed the final reflex test. But Kosch and colleagues have painstakingly preserved tissue from every last one.

'We've kept the skin sections and we're going to study the skin and see if there's anything unique,' she says. 'Then, we're going to actually develop a genotyping method from the tissues of these frogs, so not a piece of them will be wasted.'

Genotyping is a technique that can detect mutations in DNA between individuals of the same species that can lead to major changes in the phenotype. Genotyping will be crucial in identifying which genetic mutations the resistant frogs share that might be transferable.

There are all sorts of methods for genotyping, but in this frog-tissue pilot study, the team hopes to use another cutting-edge tool.

'The most common method for wildlife is really crude – you basically just digest the genome with what are known as restriction enzymes, and then you look at certain sized pieces that remain and you sequence those to look for variation,' Kosch explains. 'So, you don't really have any control over what parts of the genome you look at.'

SNP-ChIP is a bit different. SNP stands for single nucleotide polymorphism, which describes a change in the genetic code at the level of a single base pair on the genome – essentially the fundamental unit of the DNA double helix. Base pairs contain a combination of two of the four nucleobases found in DNA – either adenine and thymine, or guanine and cytosine – which form the rungs on the double helix ladder.

If a base pair is polymorphic, some members of a species may have that base pair at that location on the genome, while others will have a different base pair. It helps if you imagine the genome as a long metal chain: ten links into the chain some people might have a gold link, while others may have a silver link.

So, SNP changes are very small but very common – on average, an SNP occurs after every 1000 base pairs in the genome (for reference, the human genome contains approximately 3 billion base pairs, while the southern corroboree frog genome contains three times as many).

The SNP method is thus more refined than other traditional genotyping methods.

To identify the target SNPs on the corroboree genome, scientists introduce synthetic, fluorescent nucleobases that bind to the corresponding bases on the DNA, flagging genetic variations and potential areas of interest.

If the frogs that survive chytridiomycosis share some clear SNPs, that will signal to the research team where their hunt should focus. If those SNPs are close to genes that are known in other species to code for resistance to other pathogens, or sit close to genes known to code for an immune molecule like a T cell, that's another potential target.

Since there will likely be at least 50 000 SNPs per frog, they'll need a computer program to conduct statistical analysis to identify potential candidates. And even then, it's not perfect.

'There will be some [SNPs] that are found in the resistant frogs that have nothing to do with it, they just happen to be there,' Kosch says. 'So then you've got to try and rule that out as well.'

It's going to be a complicated and painstaking process of elimination. But if they find that mutation needle in the DNA haystack, they might be able to save the species.

Whether you're nervous about Franken-frog or not, this kind of work is likely to become more common, as we enter the adaptive phase of our reckoning with the compounding impacts of climate change, habitat loss and biodiversity collapse.

A million species are at risk of extinction, according to the UN. Most of these are disappearing because the equilibrium they evolved under is shifting. Oceans are warming, tree canopies are being replaced by arid grasses, rainfall patterns are growing or shrinking, fires are breaking their historic bounds. Change is coming. In many cases, change is already here.

So, do we owe it to the species we've imperilled to save them, by whatever means necessary? Or is this just another experiment in human hubris?

'In the face of more and more pressing threats, from climate, invasive species, other biodiversity threats, using all the powerful tools, including genetic engineering, is necessary,' argues Newhouse. Gyngell, within reason, tends to agree.

'My perspective on it is very technology neutral,' he says, 'so I don't think technologies are good or bad, it's how you use them.

'I think we can just have this mentality that we need to protect our ecosystems, and we should be using whatever tools we've got at our disposal to do that.'

For the southern corroboree, which evolved in tune with its now threatened homelands on the Snowy Mountains – an area that was almost wiped out by the 2019–20 bushfires – things are looking dire. Without a pair of molecular scissors, argues Kosch, we might not be able to save them.

'If we want these frogs to be back in the ecosystem, this is the only approach that has plausibility.'

❋ *This magnificent wetland was barren and bone-dry. Three years of rain brought it back to life*, p. **28**
  Noiseless messengers, p. **162**

A WHOLE BODY MYSTERY

Alice Klein

I was 19, my face raging with acne, when my dermatologist started asking me questions that seemed to have nothing to do with my skin. 'Are your periods regular? Do you have any excess body hair?' he asked. 'You may have polycystic ovary syndrome,' he concluded. I had no idea what he was talking about. 'It can make it difficult to have children,' he said as he saw me out.

Reeling, I went to my family doctor, who ordered blood tests and an ultrasound of my ovaries that confirmed I had polycystic ovary syndrome, or PCOS. But she admitted she didn't know much about it, leaving me confused and miserable about this mysterious condition I had suddenly been saddled with.

Many of my friends have recounted similar experiences. Despite PCOS being the most common hormonal condition among women aged 18 to 45 and a leading cause of infertility, it has been hard for us to get a straight answer about what it actually is or what to do about it.

Seventeen years on from my diagnosis, however, the tide is turning. Researchers are finally piecing together the causes of PCOS and it is being taken seriously as a condition that doesn't just affect the ovaries, but also has cardiovascular, metabolic and psychological repercussions. As a result, the condition is even set to get a different name later this year. And what's more, this clearer understanding is opening up routes to new treatments.

The first doctors to characterise PCOS were Irving Stein and Michael Leventhal at Northwestern University in Chicago. In 1935, they published a report on seven women with similar symptoms: cysts on their ovaries, irregular or no periods, unsuccessful attempts to become pregnant, and some with acne, obesity or excess hair on their faces or bodies. The condition was originally called Stein-Leventhal syndrome before later becoming known as polycystic ovary syndrome.

Today, a PCOS diagnosis is based on having two of three characteristic features. The first is high levels of male sex hormones like testosterone, which can cause acne, excess hair on the face and body, and thinning head hair. The second is irregular or no periods, which occur because eggs often haven't developed properly in the ovaries. This prevents their regular monthly release in the form of ovulation, meaning that it can take longer to become pregnant. The third is the presence of 20 or more 'cysts' on either ovary, which are now understood to be eggs that are stuck in an immature state, rather than actual cysts.

Multiple impacts

In addition to these key features, around 50 to 70 per cent of individuals with PCOS develop resistance to insulin, which can lead to higher levels of this hormone, type 2 diabetes, weight gain, high blood pressure and heart disease. PCOS also increases the risk of endometrial and pancreatic cancer, and can cause anxiety, depression and reduced sex drive in some people.

The psychological effects may be directly caused by hormonal imbalances. Alternatively, they might arise because 'if you're a teenager, when PCOS symptoms emerge, and you're gaining weight rapidly, you have significant acne, your periods are all over the place and you have body hair where you don't want it, it can have a really significant impact on your self-esteem', says Helena Teede at Monash University in Melbourne, Australia.

Finally, people with PCOS who become pregnant are more likely to have miscarriages or complications like gestational diabetes or preterm birth.

PCOS affects around 5 to 18 per cent of cis women and up to 58 per cent of trans men, although the reason why this latter figure is higher has yet to be pinned down. Despite being relatively common, it has long been one of the most neglected health conditions, says Teede. 'It's twice as common as diabetes but gets less than a hundredth of the funding,' she says. Elisabet Stener-Victorin at the Karolinska Institute in Sweden tells a similar story. 'Up until about 10 years ago, I would never put "PCOS" in the title of my research grant applications because it really dragged down my chances of getting funding,' she says.

Part of the problem is that it is 'everybody's business and nobody's business', says Teede. The many symptoms of PCOS, which vary widely between individuals, means it is managed by a range of health professionals: endocrinologists, gynaecologists, reproductive specialists, dermatologists, primary care doctors, dieticians and so on. For a long time, no one was sure who should be steering the ship and each speciality treated PCOS differently, which 'constantly created confusing messages', says Teede.

To rectify this, Teede led the development of the first international, evidence-based guidelines for PCOS, which were published in 2018. They were based on consultations with more than 3000 health professionals and people with the condition from 71 countries. 'We needed a really strong cut-through with all the experts in the world saying the same thing,' she says.

The guidelines explain how to diagnose PCOS and manage it using existing treatments. Diet and exercise interventions are recommended to begin with, since these have been shown to simultaneously improve the metabolic, reproductive and psychological features of the condition. This is because diet and exercise can assist weight loss and improve blood sugar control, which, in turn, reduce insulin and testosterone levels.

Personally, I have had some luck with lifestyle management. I tried a low GI (glycaemic index) diet after reading a small study that showed that 95 per cent of women with PCOS who adopted this diet – which involves eating foods that minimise blood sugar spikes – developed more regular periods. Amazingly, my menstrual cycles shortened from around 70 to 40 days when I tried it, but I wasn't able to keep it up long term because of my love of white rice and bread.

If lifestyle changes aren't enough, certain medications can also help. The oral contraceptive pill, for example, can regulate periods and reduce acne and unwanted body hair. A drug called isotretinoin can also ease acne – it cleared mine up in a matter of weeks – and laser treatment can remove unwanted hair. Letrozole can stimulate regular ovulation in individuals trying to conceive and metformin can help to combat insulin resistance and weight gain.

These treatments don't always work, however, and they don't get to the root causes of PCOS. 'There is no cure so far – all the treatment options available treat the symptoms and not the disease itself,' says Paolo Giacobini at the French National Institute of Health and Medical Research. He and others are now trying to develop PCOS-specific drugs.

To do this, they first need to understand exactly what drives the condition. A starting point is that it often runs in families. Stener-Victorin and her colleagues, for example, found that women in Sweden were five times more likely to be diagnosed with PCOS if their mother has the condition. No single gene has been found to be responsible for PCOS, but certain patterns of genes involved in testosterone production, ovarian function and metabolism appear to be linked with the condition. Still, these genetic variations don't tell the whole story of how PCOS is passed down generations.

Growing evidence suggests PCOS-related hormonal imbalances during pregnancy can also have an effect on the fetus. 'In a woman with PCOS, you have both the genetic factors and the in utero environment,' says Stener-Victorin. 'I think it's likely that you may carry some susceptibility genes and then you have an in utero environment that triggers its onset.' Two hormones suspected to be involved in this in utero effect are testosterone and anti-Müllerian hormone (AMH), both of which tend to be elevated in those with PCOS.

Stener-Victorin and her colleagues have found that injecting excess amounts of a form of testosterone into pregnant mice caused their female offspring to develop many of the hallmarks of human PCOS, including irregular cycles, and greater fat mass and body weight. Similarly, when Giacobini's team injected excess AMH into pregnant mice, their female offspring had irregular cycles, the appearance of 'polycystic' ovaries, elevated testosterone, insulin resistance, higher body weight and greater fat mass.

'We now have an animal model that not only recapitulates the reproductive aspects of PCOS, but also the metabolic component seen in many women,' says Giacobini. 'So, we can use these animals to really investigate the disease and design new treatment options.'

Most recently, his team discovered that the daughter mice with PCOS-like symptoms, whose mothers were injected with excess AMH during pregnancy, had altered expression of several genes involved in inflammation. This has led Giacobini to believe that PCOS is actually an inflammatory condition.

His team found increased expression of inflammatory genes in the brain, ovaries, liver and fat of the mice, which he says may explain why these organs are all affected by the condition. This fits with emerging evidence of a link between inflammation and PCOS in people. A 2021 analysis led by Saad Amer at the University of Nottingham, UK, for instance, found that women with PCOS had significantly higher levels of an inflammatory marker called C-reactive protein compared with those without the condition.

Could these findings lead to new treatments? Giacobini's team has spent the past few years developing drugs to lower AMH levels. The researchers are about to test these in mice, before hopefully progressing to human trials. 'But we need to be very cautious because there are AMH receptors in different parts of the brain and a range of organs,' he says.

'We cannot predict yet whether such treatment may trigger undesirable side effects until we fully comprehend the role of AMH in all those organs.' Interestingly, AMH declines with age, which may explain why some with PCOS who were unable to conceive naturally in their 20s and 30s are able to do so in their 40s, when their AMH levels fall into the normal fertility range, says Giacobini. This delayed fertility window could also be the reason why those with PCOS reach menopause four years later than average.

New way forward

Another treatment option may be drugs that correct the altered expression of inflammatory and other genes implicated in PCOS, says Giacobini. Last year, his team showed that PCOS-like symptoms could be reversed in female mice by giving them a drug called S-adenosylmethionine that corrected the altered gene expressions. This drug couldn't be safely given to people because it affects too many other genes, but it may be possible to develop more tailored treatments in the future, says Giacobini.

Teede says these approaches are worth pursuing, but cautions against extrapolating too far from animal studies. 'PCOS is not caused by one mechanism, it's multiple mechanisms that add up together,' she says. 'If you've got an animal model that uses one mechanism to induce a PCOS-like status, you might be able to reverse that one mechanism, but treating a complex multifactorial condition in humans is harder.'

In the meantime, Teede believes that PCOS management could be vastly improved just by providing people who are diagnosed with the condition with better information about what it is and how to manage it.

There are still many common misconceptions about PCOS that need to be addressed, she says. For example, my biggest worry when I was diagnosed was that I wouldn't be able to have children – a concern that is very common, says Teede. In fact, 'research shows that women with PCOS have the same family sizes as others. Often they just need a bit of a help', she says. 'That doesn't have to be IVF – medication that stimulates ovulation is often all that's required.'

To help bust these myths, Teede and her colleagues released a free app called AskPCOS in 2018 that provides evidence-based answers to the 93 most common questions asked about the condition. 'It's now in 12 languages and is used by about 30 000 women in 176 countries,' she says. 'It's important to have something like this because there's so much rubbish out there – people are trying to make money off vulnerable women by selling diets and supplements for PCOS that have no evidence.' At the same time, her team has created simple resources for health professionals to allow better diagnosis and management.

I ask Teede if it's time to change the name 'polycystic ovary syndrome', since it's now recognised as a whole-body condition, people can be diagnosed with it even if they don't have 'polycystic ovaries', and we now know they are not cysts. 'We desperately need a name change,' she says. 'The name should reflect what it actually is. Having a name around the ovaries misses the point and gives false implications.'

Teede and her colleagues are currently consulting health professionals and people with the condition to agree on a new name – the most preferred one at this stage is 'reproductive metabolic syndrome'. They hope to formalise this name change at the end of this year when they release an updated version of the international guidelines.

My own journey with PCOS has been unpredictable. After all those years worrying that I wouldn't be able to have a family, it was a happy surprise to conceive my two children naturally. However, there were several miscarriages along the way that may have been related to my PCOS.

The next twist came after my pregnancies, when my once erratic periods suddenly became like clockwork and have continued like that to this day. This is apparently quite common, although no one knows why.

There are many mysteries of PCOS that still need to be unravelled, but it seems like we are finally gaining a better understanding of the condition and improved diagnosis, education and treatment. I just wish I could go back to that 19-year-old girl leaving the dermatologist's in tears and tell her it was going to be all right.

✱   *Ears*, p. 71
    Long Covid: After-effect hits up to 400 000 Australians, p. **154**
    The psychedelic remedy for chronic pain, p. **195**

POINT OF VIEW

Lauren Fuge

I'm standing at the base of Lathamus Keep, watching tree climber and photographer Steve Pearce attach my harness to an orange rope no thicker than my finger. It's one of four climbing lines he and his crew have set, using light lines attached to weighted throw bags – an impressive mission when the first branch is 25 metres up. The tree's moss-patterned trunk is a wall of wood before me, twice as wide as my outstretched arms. I squint into the January-bright canopy.

'How far up are we going?' I ask.

'The line's set at 70 metres,' Pearce says, with the casual tone of a person who has spent thousands of hours aloft.

'And how tall's the tree?'

'Eighty metres. Biggest blue gum in the universe.'

Biggest blue gum (*Eucalyptus globulus*) in the universe, roughly the size of the launch structure of NASA's Artemis Moon mission, and it's just an hour and a half from nipaluna/Hobart.

'This forest is called the Grove of Giants,' explained the endlessly enthusiastic canopy ecologist Dr Jen Sanger, as she led me through the dappled dreamscape of wet eucalypt forest that morning. 'There's about 150 trees here over four metres in diameter, so it's just jam-packed with giant old trees.'

The forests of lutruwita/Tasmania are one of just three places in the world where trees grow above 80 metres. In the teeming rainforests of Borneo, yellow meranti trees soar up to 100 metres; the fog-shrouded west coast of North America creates the perfect conditions for temperate rainforest species to reach even more epic proportions; here, five species of eucalypt shoot up above the smaller rainforest trees to become islands in the sky.

This giant, Lathamus Keep, is named for the habitat it provides the endangered swift parrot (*Lathamus discolor*). Its deeply furrowed buttress resembles giant fingers driving tip-first into the soil; lanky saplings shoot up out of the gnarly metacarpals, their roots gripping onto the knuckles for structure.

It's circled by decay: dropped leaves and broken branches and strips of shed bark longer than I am tall, making the surrounding trees look like they're dripping with candlewax.

I've spent hundreds of hours walking through forests before – I consider myself a tree person. But I often ignore them. They're everywhere, every day. Yet in the kaleidoscope of information relayed to my brain every millisecond, trees usually don't make the cut. And I'm not alone: US botanists coined the term 'plant blindness' to describe this inability. It's not universal, but it's a real phenomenon for a great chunk of us humans.

Right now, I'm the least plant blind I've ever been. I attach my foot ascender to the climbing line and shift my weight off the ground. Suspended, gently spinning, it's impossible not to pay full attention to this living, breathing being. My life now depends on it.

'If you were an alien, coming to Planet Earth, the first thing you'd probably notice is that there's a lot of water, because it's blue,' says Margaret Barbour, a plant physiologist at the University of Waikato in Aotearoa New Zealand. 'And the second thing you'd notice is that on land, it's green.'

With four billion hectares of forest around the world, trees are 'the Earth's chief way of being', in the words of author Richard Powers. Some 380 million years ago, as the first tetrapods crawled up out of the swamps onto dry land, trees shot up from ankle-brushing plants into 30-metre giants. While our ancestors worked out how to breathe with primitive lungs, trees rapidly rose to the status of gods. The first forests transformed the planet, becoming the dominant drivers of the biogeochemical cycles that run all life on Earth – carbon, water, nitrogen, phosphorus.

Trees also inherited the gift of photosynthesis first developed in stromatolites 3.4 billion years ago: their leaves take in carbon dioxide and light from the atmosphere to power their growth. This not only modifies the atmosphere by releasing oxygen, but provides habitats for animals and forms the basis of everything we eat.

'That, I think, is the magic of plants – they can convert radiant energy and CO_2 into a stored energy that all life depends on,' Barbour tells me. 'That's something we learn in primary school, but it still has that magic.'

Up, up, up, into free space, watching the thick, textured bark turn smooth and creamy as I climb, feeling like I'm closing an evolutionary loop. My foot ascenders slide easily along the rope then catch, allowing me to step up as if on a rope ladder. But the act of climbing isn't easy at all: every 7 or 8 vertical metres, muscle fatigue forces me to stop and rest, swinging gently over the world.

Today, humans are the only primates that don't spend time in trees. Some 6 million years ago, our ancestors moved down onto the grassland savannahs for the first time, where their postures straightened and their knees and feet grew better suited for walking than canopy-swinging. Their world flattened, and became ours. The years ticked past; the canopies remained out of reach. Botany boomed briefly, clipping and classifying the forests Linnaean-style, then science moved onto more sensational ambitions: the deep ocean, the frozen poles, infinity and beyond.

In Richard Preston's 2007 book *The Wild Trees*, French botanist Francis Hallé recalls walking through the rainforests of New Guinea with students in the 1970s. 'We were looking at the tree canopy – so many epiphytes, so many animals,' Hallé says. 'A student said, "Funny! Man is able to collect stones on the Moon but unable to work in the canopy".'

In those early days, researchers tried to collect data using blimps, aircraft and cherry pickers with buckets to lift them up into the treetops.

'The French even designed a massive hot air balloon that has this platform that lies on top of the canopy,' explains Sanger, who came to the field in the 2010s to study Australia's subtropical rainforest canopies. 'It's got these holes in it so you can ... walk around and just stick your hand down in the hole and grab some leaves to sample.'

One of the first to use a direct climbing method – using just your body and ropes – was US scientist Meg Lowman. As a young PhD student in 1970s, Lowman came to Sydney to study tree dieback in northern New South Wales. 'She met a whole bunch of cavers, and they helped her get up into the trees,' Sanger says.

In the canopy, Lowman discovered the root cause of the dieback: surging populations of insects in response to the warming, drying climate.

In the decades since, direct climbing has become the least invasive and most economical option for canopy science, but still these trees remain hard to access, especially the temperate forest giants whose first branches are often dozens of metres up. It wasn't until the 1990s – 40 years after Hillary and Norgay clambered to Everest's summit, and a quarter of a century after Armstrong stepped onto the Moon – that US botanist Steve Sillett discovered the world of redwood canopies in California.

No one looked at the canopy biodiversity of southern Australia's giant eucalypts until the early 2000s, when US-trained masters student Yoav Daniel Bar-Ness trapped and surveyed their insect life.

Very little work has been done here since. When Sanger and her husband Pearce began documenting and measuring Tasmania's tall trees in 2015, they were among the first to return to these canopies in a scientific capacity.

And yet, our bodies still hold memories of our ancestors' arboreal lives. My hands grip the rope with opposable thumbs that stuck in our evolutionary line because they're useful for grasping branches. My eyes pick out the purple of Sanger's jumper against undulating green and brown, thanks to an extra type of light-sensitive cell useful for spotting ripe, colourful fruit – a trick of the forest to distribute its seeds. We were shaped by forests for millennia before we became *Homo sapiens*. My muscles may have forgotten how to climb, but this isn't a journey into the unknown: it's a return.

Just below Lathamus Keep's first branch, I sit back in my harness and look up: there's still an impossibly long way to go. I look down: I'm level with the very top of the rainforest understorey, a mosaic of celery-top pines and leatherwoods and tree ferns. In a moment, the forest floor will vanish beneath their interlocked crowns. In a moment, all I'll have is tree and line and sky.

When botanists first climbed a hundred metres up into the California redwoods, they found what Preston calls 'coral reefs in the air': hanging gardens of ferns and mosses and lichens, ponds in old hollows swimming with plankton, beetles digging through rotting wood, and soil forming in forks that supported fruiting huckleberry bushes and bonsai trees, all feeding on moisture from fog.

The canopies of wet eucalypt forests are entirely different. They aren't bathed in constant coastal fog, and they don't provide a stable enough habitat for epiphytes because they shed their bark like snakeskin – dislodging parasites and evicting any moss, fungi or lichen that have moved in.

Instead, these islands aloft are dominated by animals and insects. As eucalypts reach middle age they begin to form gnarled hollows, which become homes for possums and microbats, spiders and skinks, insects and a diversity of birds.

As I climb past a hollow decorated with the lacework curtains of spiderwebs, I feel a strange kinship with their maker, dangling from a fine thread in a tree many thousands of times my size.

Above the understorey, the world is made of light.

I climb through the fork of the first branch, where it cleaves in two and curves up like a living candelabra, just the first of many. After 25 metres of limblessness, my line runs in and around a woody maze that I work my way through, bumping against burls and batting away the whippy leaves of fresh shoots.

From below, perspective compressed the tree to a couple of layers. Up here, I'm clambering through endless storeys of branches – each the girth of the biggest tree in a suburban park – splitting and replicating and fanning out to catch stray rays of sunlight that filter past the dozens of limbs still above.

When I next stop to catch my breath, an intense hum drives me to distraction, vibrating directly into my skull.

Pearce has patiently been climbing at my pace, and I ask him, 'What insects are we hearing?'

'They're crickets.'

'It sounds like they're in the trees.'

He laughs. 'They're probably all around us, mate. It's a three-dimensional space.'

For most of my life I've been moving through the world in a rigid XY plane, and now I've flipped into another dimension – the Z-axis. But while I can only travel vertically, every other critter can move in the fullness of three dimensions.

'There's 30 or 40 metres of forest below us, and there's still 40 metres of forest above us,' Pearce says. 'Different species and life forms occupy different zones in that space.'

In springtime, he tells me, birdsong contours the 3D world.

'From the ground, everything's above you. But when you're in the middle, it's like – the thornbills only operate down there, and the grey fantails work there and that's their zone … and then there's the eagles and the cockatoos at the very, very top.'

I'd thought climbing would give me the perspective to appreciate the size and significance of this tree. But just like on the ground, inside its convolutions it's impossible to see the whole.

'The way we often think about organisms and environments – and this is despite our best intentions – is to think about an organism as something that develops externally to its environment,' says Dalia Nassar, an environmental philosopher at the University of Sydney.

And yet, everything on this planet has developed within intricately interconnected life support systems: the lithosphere, hydrosphere, biosphere, atmosphere.

'The organism is a part of the environment in such a way that it's almost impossible to separate it. You can't say where the organism ends and where environment begins.'

Plants, Nassar points out, demonstrate this far more radically than animals.

'Plants are rooted, they are of their environment, they are of the soil – they are transformed by the soil at the same time that they are transforming the soil,' she says. 'They are determining the temperature, they are determining the rain season, just as much as they are determined by it.'

Much of our scientific understanding of tree–environment relationships has been revealed in recent decades, according to the University of Waikato's Barbour, who works with Nassar to create a conversation between plant science and philosophy.

'There's a whole lot of things we've learned about how plants respond very, very sensitively to their environment,' Barbour explains. 'Because trees can't move, they can't move away … when things get rough. They have to be able to respond to environmental stresses.'

We've known about how they respond to light for a long time, she says. 'But more subtle things like nutrient availability and water availability within the soil and how that varies over space and time, and how plants respond to those variabilities – that's the kind of thing that we're just learning about more now.'

Trees of the same species can grow differently according to how much light and water they receive, or how densely they're planted, or the wind or soil conditions. Even in an individual tree, the leaves at the lighter end of the canopy are anatomically distinct from the leaves below.

'The tree senses its context from the beginning and develops in dialogue with it,' Barbour and Nassar write in an *Aeon* essay exploring their collaboration. 'Trees are so adaptive to their surroundings that a human equivalent to tree plasticity would be certain people growing large webbed feet (like diving flippers) simply because they swim a lot.'

In turn, trees engineer their world as they grow, both locally – creating microenvironments that affect light, nutrients and water, often determining which species grow around them – and globally, driving biogeochemical cycles.

There's also evidence under some conditions for plants picking up on the chemical signals of others – say, the volatiles released by a tree attacked by insects – as well as emerging research that shows some species share resources through underground mycelial networks.

Trees aren't static or passive beings: they're active agents in their environment, both shaping the world and being shaped by it. In fact, Barbour and Nassar argue that trees are synecdoches – 'a part that signifies or expresses a whole'.

'We have, for too long, seen natural beings as static objects that develop in separation of one another,' Nassar tells me. 'We have failed to properly conceptualise the dynamic and collaborative nature of the world.'

In reality, we're all dynamic processes, developing in collaboration – humans included.

'As living beings, we're participating in these processes,' Nassar says. 'We're affecting them as much as they are affecting us. Like plants, we're transforming the environment in vast ways, but we have to think about how this is transforming us.'

The crown of Lathamus Keep doesn't peter out: it stretches expansively into a multi-tiered dome, big enough to hold dozens of people in its branches.

I stop climbing just before the end of my line, when I find Sanger lounging in a massive fork as if it's a hammock. The other climbers – Sophie and Kevin and Ethan – are already here, and we join them, draping ourselves over branches or hanging suspended in space, the six of us, 70 metres aloft.

Peanut-butter-and-jelly muesli bars are passed around; we point out blossoms and talk about government logging quotas and tell stories of the day they measured this tree to be the biggest blue gum in the world, sending a climber up to its topmost branches, no thicker than my wrist.

As we chat, I try to think of all the things I'm not seeing: the millions of processes and interactions invisible to the human eye. This organism is pulling water 80 metres up using an impressive vascular system. Every leaf in sight is using this water as it takes in carbon dioxide through its stomata and produces energy through photosynthesis. This magic of light into life continuously builds carbon into the skin of Lathamus Keep, into its roots and trunk and branches.

This forest, Sanger tells me, is incredibly carbon dense. With a team of citizen scientists, she recently surveyed two hectares of the Grove of Giants – climbing and measuring the eucalypts, the understorey and the soil. Preliminary results show this forest holds the highest amount of carbon of any forest in Tasmania.

'There's about 1250 tonnes of carbon per hectare in these forests, which is an absolutely phenomenal amount,' she says. 'Globally, that's some of the highest carbon [density] you'll ever find.'

In the process of building itself layer by layer, Lathamus Keep is also producing a gift of its own: oxygen. Just like us, every plant on this planet is constantly breathing in and out, creating and maintaining the composition of the thin blue skin of an atmosphere we all depend on.

'They're the reason we're here to begin with,' Sanger says.

I know this intellectually, but every time I see a tree, I don't think: *you're the reason I'm alive*. I get too caught up in my own bubble to think of all the intimate chloroplastic relations of our planetary one. But here in the forests of southern Tasmania, here with my whole weight supported by this living pillar of carbon and water and sunlight, I'm beginning to see how much our lives depend on it.

Let's leave me dangling in the crown for a moment. Bear with me; we're going higher.

In March 1965, Soviet cosmonaut Alexei Leonov became the first human to walk in space. He hurtled into orbit in the tiny Voskhod 2 spacecraft, crammed up against fellow cosmonaut Pavel Belyayev. As they drifted above Egypt, Leonov suited up, pulled himself into the void, and let go. A thin line was all that tethered him to life. For 12 minutes he spun gently as the Earth spun beneath him – bright and curved, with clouds and mountain ranges and oceans and continents speeding by.

'The Earth was small, light blue, and so touchingly alone,' Leonov later wrote. 'Looking back at our blue globe from such a distance profoundly changed my vision of space and time.'

Many who have ventured into space since have recorded a similar feeling: a kind of orbital perspective, gained from seeing our world from the outside. Hanging in the star-filled void, a switch seems to flip in their brains and they can see the interconnectedness of our planet and its systems – see its vastness and its insignificance, its wondrousness and its fragility.

Edgar Mitchell, lunar module pilot on Apollo 14 and the sixth man to walk on the Moon, explained to *People* magazine in 1974: 'You develop an instant global consciousness, a people orientation, an intense dissatisfaction with the state of the world, and a compulsion to do something about it. From out there on the Moon, international politics look so petty. You want to grab a politician by the scruff of the neck and drag him a quarter of a million miles out and say, "Look at that, you son of a bitch".'

This phenomenon has been termed the 'overview effect'.

It was profoundly difficult for the first astronauts to bring what they saw home. But Leonov was an artist.

After the unexpectedly harrowing experience of returning through the tiny airlock – his spacesuit had inflated, forcing him to manually vent oxygen – an exhausted Leonov reached for his sketchpad.

Weightless, with coloured pencils attached to his wrist by string, Leonov sketched his impressions of the spacewalk. But he didn't draw Voskhod 2, or himself floating in space, or even the endless vistas of stars. Instead, this first astral artwork depicts the fragile blue curve of our atmosphere pressing up against the void, with the sun alight in the threshold between.

'I tried to capture the different shades of charcoal rings that make up the Earth's atmosphere, the sunrise or air glow over the Earth's horizon, the blue belt covering the Earth's crust, and the spectrum of colours I had observed looking down at the globe,' Leonov later said.

And when he returned home, this sketch came with him – bringing us our first vision of the Earth from above.

When Pearce suggests that we start heading down, I don't want to go.

Up in the crown I have a sense of infinite space: from the shadowed understorey I've stepped up into a place of vast horizons, punctuated only by other giant trees, each its own island in a wrinkled sea of rainforest canopy.

It radiates out for 100 hectares, and beyond I see waves of hills that have been logged and uniformly replanted – severing the complex, long-tended relationships that intact forests cultivate with the earth and sky. This grove is within the logging coupe, and may be next.

On the 2D surface of the planet, we're safe within our own perspectives of the world. But like launching into space, ascending into the canopy adds another dimension. Suspended here by a single tether, looking out and up, I'm confronted with how intimately and integrally we're tied to this place; I can see what we're losing by cutting our own lines.

I unclip my foot ascenders so I'm supported just by my harness. My climbing line drops off into the abyss behind me, falling some 70 metres to a ground I can't yet see. The only thing holding me up is a single device gripping the line.

I pull the release lever, and begin the journey home.

Some 20 kilometres south of the Grove of Giants, within the Tasmanian Wilderness World Heritage Area, is another incredibly carbon dense forest.

'These tall eucalypt forests are world-renowned for the amount of carbon they can store,' says Tim Wardlaw, forestry scientist at the University of Tasmania and site manager of the Warra SuperSite.

He would know, because he can measure it.

Warra is the southernmost node of the Terrestrial Ecosystem Research Network, and it's equipped with an 80-metre instrument tower capable of mapping the flow of water, nutrients and energy through the forest of messmate stringybark (*Eucalyptus obliqua*).

The tower was only installed about a decade ago, but the world has changed a lot since then.

Evidence had previously shown that forests respond to a heatwave in one of two ways. If there isn't much water, they shut their stomata to stop water loss, and thus take in and store less carbon, becoming less 'productive'. If water is still available, then the stomata stay open and trees become more productive, as photosynthesis is efficient at high temperatures.

These two scenarios were 'the status quo', Wardlaw says: 'All the models that are in existence in Australia assume that behaviour. And then Warra came along.'

November 2017 brought three weeks of hot weather. The soil and the atmosphere weren't particularly dry, so the trees didn't shut their stomata. But the tower's measurements during the heatwave showed that photosynthesis virtually stopped.

'This is the first time it's ever been seen,' Wardlaw says.

In fact, Warra became a net source of carbon, putting about a tonne of carbon per hectare back into the atmosphere – a massive number when extrapolated out across the 800 000 hectares of forest around the site.

'That's the flipside of being a really productive forest,' he says. 'It can flip from being a really strong carbon sink to a really strong carbon source.'

The exact mechanism is not yet understood, but Wardlaw says they need to find out fast. If forests stop absorbing and start emitting carbon as the climate heats up, this could initiate a strong positive feedback loop that both exacerbates warming and endangers these forests.

'The question then becomes, can the forest acclimate?' Wardlaw asks.

And what will happen to us if it doesn't?

Down, down, down, into free space, abseiling back to the world of tarmac and currency and nation-states, the world we're building at the expense of the living one.

Down, down, down, we've dug, to the bodies of long-dead trees and plants and organisms that were transformed into coal seams, oil deposits and gas reserves by the immense pressures of rock, heat and deep time. Millennia-long cycles submerged these dark rivers of carbon and energy. By forcing our way down into extinct worlds, we've cheated the system.

'Like exorcists,' writes Andri Snær Magnason in his 2020 book *On Time and Water*, 'we disturbed their infinite sleep, pumping them back to the surface, rekindling fires and harnessing hundred-million-year-old sunshine as it lay dormant in the Earth's belly.'

As we burn these fossil gods we release their ancient carbon. The world warms. Today's forests, evolved to drink in carbon, struggle to absorb the excess. Some will switch to emitting it. The world will warm further, driving more intense and more frequent fires, heatwaves and droughts, pulling us deeper into the feedback loops that we're fighting to untangle ourselves from.

'We live,' Magnason writes, 'in mythological time.'

We've become mythological beings, capable of modifying physical and biogeochemical systems at a planetary scale. We're busy breaking cycles even as we try to remember that we're part of them – that we've been in a symbiotic relationship with this planet since long before tetrapods took their first breaths.

Trees are mythological beings too, capable of transforming the world in just as extraordinary ways. Research is now revealing how entirely they are transformed in return – they're of the soil and yet create it; they make the atmosphere and yet it literally makes and breaks them.

Like us, they're animated carbon, alight with the endless possibilities of the world. They're inescapably bound to the planet, and so are we.

As I dip below the canopy, a bright breeze turns to shadowy stillness. In moments, my feet are once again on the spongy forest floor beside Lathamus Keep. After the long, tiring, stop-start ascent, it took just minutes to drop 70 metres down.

I take stock: I've lost skin on my palm, scratched my arms, exhausted my shoulders – all gentle reminders of the gift of the climb, of existing alongside this living, life-giving being.

As everyone takes off their helmets and harnesses, we joke and chatter along with the birds. Lunch is the main topic of conversation. We're all starving.

We pack up the gear quickly because there's still 2 kilometres to walk, through a forest that's co-creating our atmosphere.

We head back on a 2D plane to the world we've made. As we walk, each breath is a reminder that part of us is being made and remade over and over again, high above in that open field of light.

✱ *Where giants live*, p. **187**
Talara'tingi, p. **193**

IN THE SHADOW OF THE FENCE

Zoe Kean

On most maps of Australia, lines crisscross the continent marking the boundaries of states and territories.

But on the ground, a different border has a much more immediate impact on the human and animal inhabitants of the outback.

The dingo fence starts in the green fields of Queensland's Darling Downs and stretches through New South Wales and South Australia before it abruptly ends on a high cliff's edge above the Great Australian Bight.

It traverses the traditional lands of 23 language groups, over Channel Country, scrub land and deserts.

It is more than 5600 kilometres long. If you know what to look for, you can see its effects from space.

'It's longer than the Great Wall of China, but not as well built,' says ecologist Mike Letnic. 'Its purpose is to keep dingoes out.'

A brief history of dingoes

The ancestors of today's dingoes arrived on the Australian continent between 3000 and 5000 years ago – most likely with people from Asia who travelled over the ocean on water craft.

In fact, the very existence of dingoes hints at a distant past of human contact between Asia and the Australian continent, long before the dates in our history books.

The nature of those interactions is a mystery. But the new

species stayed and became important to life and culture for many Aboriginal peoples.

The new 'top dog' was disruptive at first. Besides humans, dingoes were the largest predator on the land and are suspected of contributing to the thylacine's disappearance from mainland Australia.

However, over millennia, the environment adapted to the dingo just as the dingo adapted to Australia's deserts, grasslands, forests and beaches. They are now a vital part of the ecosystem.

British colonisation in the 18th century brought a new and tasty target for the now-established dingoes: sheep. The freshly minted colony rode on the sheep's back and dingoes became an enemy to farmers and graziers. As farming spread across the country, so did fences.

Rabbit-proof barriers sprang up to stop the spread of that small, hungry, invasive species. Those structures also did a great job of deterring dingoes.

In the early 20th century, sheep graziers proposed an ambitious scheme to keep their flocks safe: the barriers would be joined to create a single, long fence that would keep dingoes out of sheep country altogether. By the 1950s, around a third of the continent was ringed by the wire fence.

Dingoes were not tolerated inside the fence. To this day, graziers in New South Wales are legally obligated to exterminate dingoes on their lease. Within the fence and a buffer zone around it, dingoes are routinely shot and poisoned.

The fence varies in height, but much of it reaches 1.7 metres. Although dingoes can scale it, they tend not to. Instead, they trot along the barrier looking for weak spots or holes, a quest that often leads to lethal encounters with poison baits or traps.

Every effort is made to keep dingoes out of sheep grazing land, much of which is arid and remote. Although a few dingoes remain within its bounds, the fence has achieved its aim. Dingoes are kept at bay.

Cascading effects

Removing dingoes has changed life inside the fence significantly – and not just for the sheep and their owners.

When it rains, the land on the dingo side of the fence stays greener for longer. Dingo country is more biodiverse and has more small native mammals. Even the sand dunes are differently shaped on either side of the barrier.

For 20 years, Mike Letnic of the University of New South Wales has been returning to study sites on both sides of the fence, trying to unpick exactly how the absence of dingoes has led to these differences.

'The dingo fence has been a remarkable natural experiment into understanding the effects that apex predators have on ecosystems,' he says. 'Dingoes have not been present in great numbers in New South Wales for at least 80 years. And you can see the differences everywhere.'

Those differences begin with animals that are comparatively easy to spot: there are 'many, many more kangaroos' inside the fence, Professor Letnic says.

That's because, as well as having a taste for sheep, dingoes love to hunt kangaroos. Fewer dingoes means more roos.

When times are good, roo populations boom, but when the rains dry up, they face mass starvation. With no large four-legged predators to worry about, kangaroos also have time to nibble away at sensitive plants. Inside the fence, this has led to woody shrubs dominating over the diverse array of desert plants that are more quickly gobbled up.

And it's not just kangaroo numbers that are boosted by the absence of dingoes: inside the fence, feral cats and foxes prowl in larger numbers. These introduced predators played a large role in hunting 29 species of small native mammal to extinction.

Professor Letnic says the problem is particularly bad in desert areas and even worse inside the fence, where there are few dingoes to keep the cats and foxes in check.

'One of the things that makes me really sad when I come out to places like this in the desert is that I know that I've only got the opportunity to see a small fraction of the animals that once lived here.'

The small mammal species that have survived the past 200 years are far more common on the dingo side of the fence. These include species like the seed-eating hopping mouse and the mulgara – a tiny but ferocious predator in its own right.

While dingoes may occasionally snack on these creatures, Professor Letnic says their influence on cat and fox numbers makes a bigger difference, improving the environment for small mammals.

And that, in turn, makes a difference to the vegetation – and the sand dunes themselves. Outside the fence in dingo country, Professor Letnic says, hopping mice eat shrub seeds and seedlings, keeping shrub numbers down.

'Without the shrubs, there is more movement of sand, and we get much more open environments.'

Deakin University ecologist Euan Ritchie agrees that removing dingoes has profoundly altered the ecosystem.

'A lot of ecologists have deep concerns about the environmental impact of the fence,' Professor Ritchie says. He describes the fence as 'far from a perfect experiment' but says the differences either side of the fence are 'relatively convincing evidence that when you lower dingo presence or abundance, there is an impact on the ecosystem'.

'There is a war going on'

While the environmental repercussions of the dingo fence are becoming more understood, it does not alter the fact that started it all: it's hard to grow sheep when dingoes are about.

This reality makes many people in farming communities hostile to the predators – and in areas where grazing and dingoes overlap, the landscape is peppered with 'dingo trees'.

'Dingoes are hung in the tree to let people know that there is a war going on,' Professor Letnic says. 'Government agencies [and] farmers often call them wild dogs, and people think they're just kelpies or cattle dogs or Labradors gone wild.'

But Professor Letnic routinely clips tissue samples from the hanging animals to get their DNA sequenced. His verdict? 'It's dingoes.'

'They call them wild dogs because it's easier, from a public relations perspective, to kill wild dogs ... But those domestic animals just don't have what it takes to survive in the wild.'

An uncertain future

So, what to do?

Many already-arid areas enclosed by the fence are fast becoming too hot for sheep grazing, says Justine Philip, a research fellow at Birmingham University in the UK. This means much of the fence may soon become redundant.

But at the moment, it is hard to do anything except graze sheep on the dry country inside the fence. These places are mostly Crown land, Dr Philip explains, and you must be a grazier to take up a lease there. Sheep grazing is one of the only sources of income 'because that's the only thing that's supported by the government'.

Another income source is maintaining the fence itself. It's collectively funded by graziers, local and state governments, and upkeep costs an estimated $10 million per year.

Dr Philip hopes governments will help communities transition from sheep grazing, especially where climate change is starting to make it impossible. Where grazing continues, 'there are solutions at hand', according to Professor Ritchie.

'We can maintain top predators in the landscape and choose to maintain livestock production as well.'

He says a breed of guardian sheep dog called the Maremmano has successfully protected sheep in Queensland.

'And we can have small areas of fencing where you might bring livestock when they are birthing or calving.'

Meanwhile, new large-scale fences designed to keep dingoes out of rangelands are being built in Queensland and Western Australia, and Professor Ritchie is deeply concerned about what that means for the environment.

'It's just creating all these barriers to wildlife all over large areas of Australia ... It's a pretty horrible vision for the future.'

Professor Letnic says it's about striking a balance. 'It's important to remember that dingoes are a pest to livestock producers,' he says. But spaces for dingoes are important too.

'We're coming to realise that dingoes can play an important ecological role. I think it's really important that we think about places where we can keep dingoes – and maintain these healthy ecosystems.'

✱ *Noiseless messengers*, p. **162**
   'Gut-wrenching and infuriating': Why Australia is the world
      leader in mammal extinctions, and what to
      do about it, p. **220**

THE TORRES STRAIT ISLANDER ELDERS LAWYERING UP TO STOP THEIR HOMES FROM SINKING

Miki Perkins

When a baby is born on the tiny islands of Saibai and Boigu in the Torres Strait, its parents wait until the umbilical cord has dried and fallen off, then they bury the cord under a young tree such as a *woerakar*, a native hibiscus, or a broad-leafed sea almond.

This cord tree is a living symbol of the child's place in their clan and their connection to *ailan kastom*, or island custom. *Ailan kastom* tethers the child to the ancestral beliefs that connect Torres Strait Islanders to the land, sea and skies, and are passed from one generation to the next. Community elder Uncle Paul Kabai, 54, is explaining all this to me as we walk through the front garden of his weatherboard Queenslander on the remote island of Saibai. The carefully tended patch is full of bushes covered in red flowers. 'It's a special thing, it gives you identification,' says Kabai.

His home has sweeping views across a milky-blue stretch of sea to the thickly forested shores of Australia's nearest neighbour, Papua New Guinea. PNG can be reached via a 10-minute trip in a fast-moving tinnie from Saibai, our second-most northerly island. Residents in this region of the Torres Strait have the right to travel passport-free between the two nations, in recognition of their long history of familial and trade connections.

Kabai's own family tree grows next door at his brother's house, a giant sea almond with a small hollow in the roots, where the cords of generations have been buried for as long as anyone can remember.

A father of eight as well as a grandfather, Kabai had thought the tree would remain where it is for generations to come. Now he's less certain.

As sunlight shines on the ocean, the wind soughs through coconut palms and a passing crocodile sends idle ripples across the bay (more on this adventurous guy later), it's easy to assume Saibai's residents live in a peaceful idyll. But in the monsoon season between December and February, it's a different story. At this time, more frequent storms are adding torque to the waves that beat against the low-lying coastline, particularly when they coincide with high tides that come in about monthly, or the twice-yearly king tides.

During these storms, seawater flows under homes, erodes graves in the cemetery, ruins soil in food gardens with salt, and floods the sewerage networks, increasing the risk of disease. Saibai's $24 million, federally funded sea wall, made of boulders and concrete, was breached less than six months after its 2017 completion, causing significant damage. A $15 million, 1-kilometre sea wall on Boigu, Australia's northernmost point, was finished this year. Many wonder how long it will take for it to be breached, too.

Torres Strait – also known as 'Zenadth Kes', an amalgamation of local language names for the four winds that pass through the region – encompasses at least 274 islands in the shallow tidal waters between the Cape York Peninsula and PNG. Its population of about 4500 live on 17 islands, some as little as about a metre above sea level.

The climate crisis is a stark reality here, where residents face an urgent, existential threat to their way of life. It's one thing to understand this theoretically, quite another to feel it on the ground, in homes that have seawater sluicing underneath them.

Such rising seas are caused by rapid global warming, which is melting vast chunks of ice sheets at the poles. Nobody recalls when the changes became noticeable on these islands, nor is it possible to predict exactly when they will be inundated if the current trajectory continues. But the implications are clear.

'Becoming climate refugees means losing everything: our homes, our culture, our stories and our identity,' says Kabai, shaking his head. 'If you take away our homelands, we don't know who we are.' A stocky man, who often gestures with a pointing finger as he talks – at the blue sky, at the sparkling sea, at his island – Kabai has an endearing habit of slipping his glasses up onto his cap and peering intently at me when he wants to make a point, as if to say *this is urgent, listen up*. 'If we have to relocate, it will be very sad for us. Very sad.'

Kabai and his brother-in-law, the 52-year-old elder Uncle Pabai Pabai, who lives on Boigu, are suing the Australian government. They are seeking orders that require the Commonwealth 'take reasonable care' to protect Torres Strait Islanders and their *ailan kastom* from the harm caused by climate change – in this case by setting emissions reduction targets consistent with the best available science. They're also seeking damages for the changes they say are already undermining their way of life.

'I left school in 1989 and moved to Cairns. When I came back to Boigu a few years later, everything was the same as when I'd left,' says Pabai. 'But by the mid-2000s, I could see that things were starting to change. Our culture is built on knowledge of how everything should be. So if something happens to the seas, or the winds, or the fish, we notice.'

Launched in the Federal Court in October last year, theirs is the first climate class action to be brought on behalf of Australian First Nations people, and the first to claim that the government's failure to significantly reduce emissions will force them to become refugees. 'We have a cultural responsibility to make sure that it doesn't happen, and to protect our country and our communities,' says Pabai.

Their court action, or *Pabai Pabai v Commonwealth of Australia* as it's officially known – the Pabai case, informally – is modelled on watershed action in which the Netherlands-based Urgenda Foundation backed 886 people to successfully sue the

Dutch government in 2015. Upheld on appeal, the ruling ordered that by the end of 2020, the government cut greenhouse gas emissions by at least 25 per cent of 1990 levels, which led to the rapid closure of one coal-fired power station and billions of euros in renewable-energy subsidies and investment. The outcome forged a path for other climate cases around the world, and Urgenda is advising the Australian lawyers on the Pabai case.

In a trial due to begin in June 2023, the two men will tell the Federal Court that as the ocean rises, it will flood irreplaceable cultural heritage and burial sites, including places that contain human remains or have a spiritual significance akin to the Christian heaven. In an interim court hearing in mid-July, Justice Debra Mortimer said there was 'no denying the unremitting march of the sea onto the islands of the Torres Strait', the reality being that its people 'risk losing their way of life, their homes, their gardens, the resources of the sea on which they have always depended and the graves of their ancestors'.

She went on to note, however, that 'whether the Commonwealth has legal responsibility for that reality, as the applicants allege in this proceeding, is a different question', one that nevertheless gave the case 'some considerable urgency'.

After the case was launched, the Coalition lost power in May's so-called 'climate election'. But the litigants are pressing ahead, saying the Albanese government's target to reduce emissions to 43 per cent below 2005 levels by 2030 – against the Coalition's 26–28 per cent cut – isn't enough. 'We are saying to the government, "Stop burning coal and gas",' says Pabai. 'We need to stop that. If this goes on forever this water will go up. The only way to stop it is to stop burning.'

As the eight-seater plane makes the 20-minute trip between Boigu and Saibai, the low-lying islands look like fishing nets cast across the blue Arafura Sea; a filigree of green vegetation and mangroves, and the silver flash of pools and rivers spread across a landscape that looks more water than land. The cluster of roofs and roads in the small townships on the higher, inhabited land only take up a sliver of space, hemmed in between sea and swampy hinterland. Formed by the silt that washed down nearby PNG rivers, they lie anywhere from 1 metre to 1.7 metres above sea level.

Kabai meets us at the tiny airport, an open-air room adorned with a large white sculpture of the *dhari*, or headdress, that features on the Torres Strait flag. Pabai has flown in, too. Kabai beams in welcome – most visitors from the mainland need three flights to travel here – and takes us to his four-wheel drive, parked outside. On the dashboard sit a small plastic crocodile and a toy dog that symbolise his two totems, inherited from his parents.

We rumble out of the airport and down the dusty main road. As we crawl through the small township, the men call out in greeting to people walking or sitting on verandahs, pausing to exchange news. In this small community everyone is a friend, often a relative, and they're all talking about the drama of the previous evening, when a crocodile (apparently the same one I saw earlier) crawled over the sea wall and waddled down the main road, until someone scooped it up in the bucket of a digger and plopped it back into the sea. Videos of the hapless croc have spread like wildfire on social media, and Kabai stops the car to show us its claw marks in the mud.

One of the most pressing reasons to stop the rising sea lies in a sunny glade, a 5-minute drive from the township on a dirt track that winds through thick tropical scrub. In the cemetery, a willie wagtail perches on a tombstone and flips its tail from side to side as it tilts its head, a beady eye studying us. Kabai's brother is buried here. A low seawall separates the cemetery from the coarse sand beach and a dense forest of twisted mangroves, their wrist-thick roots plunged deep into black mud at the fringe of the island.

During king tides, the sea rises up against this wall and floods into the cemetery. The wooden frames of graves have been shifted and dislodged, leaving distraught families to hazard a guess at where their ancestor's remains lie under the soil, so they can build a new, concrete grave above. One gravestone, at least 100 years old, has toppled forward onto the beach because the sea has eaten away at the sandy bank where it stood.

Kabai and Pabai talk about what it was like to grow up on Saibai and Boigu. 'We had a really good life. We were taught by our grandparents and parents how to become a man, and get a strong hold of culture,' says Pabai. 'They would give you an idea of how to live with your families, your responsibilities. We sat with them, and they talked to us.'

Each day, children would follow their parents to the family garden, inland from the village, and learn how to grow plants like cassava, taro, sugar cane and bananas. The boys would be taken to sea on outrigger canoes to learn how to hunt *waru* (turtle) and dugong. They would build a tall platform called a *nath* from mango wood and wait there, spears poised, until a grazing dugong swam underneath, ploughing a slow path through the seagrass beds.

When children reached high school they had to move to Thursday Island, the administrative centre of the Torres Strait, or the mainland, where many stayed for work or other opportunities. There was always a sense they could return, that their island home would be waiting. That's gone. But while younger generations are less likely to return, those already here are not inclined to move. 'The elders won't move away from the island. Their church is here and our cemetery is here. They can't leave their ancestors,' says Kabai. 'If the island sinks, the elders will stay with it.'

In Australia, access to justice is often limited to those who can afford it: it's prohibitively expensive to bring a case to court and if you lose, in most jurisdictions you're faced with paying the other side's costs. In the Pabai case, that's where the Grata Fund comes in. Unless you move in legal circles, you're unlikely to have heard of Grata, named after Grata Flos Matilda Greig, the first woman to practise law in Australia, in 1905.

Australian lawyer Isabelle Reinecke founded the not-for-profit Grata in 2015 to develop, fund and build campaigns around high-impact public-interest litigation. So far, there hasn't been much need for it to cover costs: 14 of the 15 cases backed by Grata have been won, including one brought by First Nations plaintiffs from Santa Teresa, south-east of Alice Springs, in a case against the Northern Territory government over dilapidated public housing.

After the Urgenda Foundation's success in the Netherlands, Reinecke and her team wondered if something similar could work here. They talked to several communities in different parts of Australia to see if any were interested in bringing a climate change case. At the same time, Pabai and Kabai had been looking for a way to protect their communities from climate change.

'When we first met Uncle Pabai and Uncle Paul [in 2021], it was obvious how compelling their case was,' says Grata's acting executive director, Maria Nawaz. 'They spoke so powerfully about the harm that climate change was already doing, and about what they'd lose if things got much worse. We knew immediately we had to support them to have their day in court.'

Respected elders in their community, both men have given their time to local councils and representative groups, but neither had previously been an activist or involved with legal action. The duo sought advice from their elders, as well as the island councils and native title groups, on whether they should lead the case. Everyone encouraged them to get involved. 'People have been very supportive, they keep asking what's happening next,' Pabai says.

The Torres Strait has a proud history of activism: in 1936, hundreds of its maritime workers and pearl-shell divers waged a months-long strike to protest harsh treatment and unfair labour conditions, forcing the Queensland government to make changes including granting them more autonomy. In 1992, the High Court overturned the colonial doctrine of terra nullius, or 'land belonging to no one', in the famous case brought by Mer Island's Eddie Mabo. And more recently, in 2020, the child-rearing practices of Torres

Strait Islanders were recognised with Queensland legislation that enshrined their longstanding cultural adoption practices.

The Pabai case is part of a global wave of climate litigation that gained ground in the wake of the 2015 Paris Agreement, which committed signatories to limiting warming. A 2022 snapshot of litigation found that the cumulative number of climate-change-related cases has more than doubled since 2015 to more than 2000 in total, with about one-quarter filed in 2020–22. Australia is a climate litigation hotspot, second only to the United States. Last year, there were 46 climate-change judgments or new proceedings filed in Australia, up from 31 filed in 2020 and 27 in 2019, according to an analysis by Melbourne Climate Futures.

Early climate cases often involved challenging proposed projects, like a controversial coal mine or power station. But there's an increasing focus on strategic litigation that seeks greater corporate or government accountability. The most high-profile recent example was that of the eight Australian teenagers who were part of the School Strike for Climate movement, who filed a class action against then-environment minister Sussan Ley, seeking to halt approval of the expansion of a coal mine in regional New South Wales.

Initially, a Federal Court judge rejected their claim for an injunction but found the minister had a 'duty of care' to Australian children to protect them from harm when assessing fossil fuel projects. The full bench of the Federal Court later unanimously overturned this decision on appeal, deciding it was not the place of the court to set policies on climate change but rather, the job of elected representatives.

In court, Pabai and Kabai will be represented pro bono by law firm Phi Finney McDonald (PFM).

I meet senior associate Grahame Best in the firm's airy office on Flinders Lane in the heart of Melbourne's CBD. We sit at a large conference table as the 36-year-old solicitor, dressed in a suit jacket and jeans, explains that a team of four lawyers had to initially examine whether they could win the case by 'working to ground' different legal theories.

'It's very strategic. You have to figure out all the different pathways and goals and there's no playbook,' he says. PFM has a strong social justice practice, including representing First Nations, refugees and low-income people. It has acted for a Djab Wurrung elder contesting the destruction of sacred birthing trees in country Victoria, and for asylum seekers detained on Manus Island and Nauru.

Best and his team will try to convince the Federal Court to establish a novel duty of care that doesn't yet exist in the law. To be successful, they will need to prove that a special relationship exists. 'The Commonwealth has a heightened relationship with Torres Strait Islanders,' argues Best. 'This is because there is a treaty in place, the Torres Strait Treaty, which sets out a series of obligations for the Commonwealth to protect and preserve the land and culture of the Torres Strait.'

In April 2022, the Morrison government acknowledged that Torres Strait Islanders are vulnerable to some effects of climate change, such as rising sea levels, and that the region has already been impacted. It denied that the Commonwealth owes the two men the duty of care alleged, however, and noted that Australia contributes only a very small proportion of global emissions. New Attorney-General Mark Dreyfus has since said he is unable to comment while the matter is before the courts.

The world has heated by 1.1 degrees Celsius since 1850, and Australia by an average 1.2–1.68 degrees since 1910. Most climate scientists say that Australia should set an emissions cut of about 75 per cent by 2030 if the country is to have any chance of doing its part in the global – and likely unsuccessful – push to limit warming to 1.5 degrees.

According to the Intergovernmental Panel on Climate

Change, even if warming is limited to well below 2 degrees, global sea levels could rise by 2100 by 30 to 60 centimetres, but by as much as 1.1 metres if emissions continue to strongly increase. With some Torres Strait islands only a metre or so above sea level, this would make them uninhabitable during storm surges.

Sea-level rise is complex to model. The Torres Strait is particularly dynamic because the powerful Pacific and Indian oceans move through a narrow channel, and the waters mix together in 'baffling' tides in the Timor and Arafura seas, says David Kennedy, a professor in coastal geomorphology at the University of Melbourne. While small and low-lying islands are sensitive to myriad impacts of climate change, it's unclear what kind of increase would be produced by warming of 1.5 to 2 degrees.

What we do know, however, is that the Australian Bureau of Meteorology has measured the rise in the Torres Strait as at least 6 millimetres a year between 1993 and 2019, or almost 16 centimetres in total. A tide-gauge station at Thursday Island, which began operating in April 2015, suggests a trend of 16 millimetres a year, although this may be due to natural variability over a short observation period.

Since the Commonwealth's initial response to the Pabai case, there has been a change of federal government. One of the first things Chris Bowen did after becoming Minister for Climate Change and Energy was travel to the Torres Strait. In June, he met with community and council leaders to hear how they were dealing with what he calls a 'real and substantial threat'. He says, 'The trip was about listening to the people of the Torres Strait, to hear from them first-hand about their experiences', noting he's unable to talk about the specific case. 'We don't have a second to waste.'

Grahame Best and his team have been comparing notes with Urgenda's legal counsel in the Netherlands, Dennis Van Berkel. The foundation was established to stimulate an environmentally sustainable economy, and introduced the first collective purchasing agreement for solar panels in Europe. Its landmark 2015 win shifted debate in the Dutch parliament, Van Berkel tells me. 'It has really changed the political landscape with regards to how politicians are dealing with climate change,' he says. 'They just know we have to deal with it. We can't get away with it anymore. We can't ignore it anymore.'

The Dutch parliament voted in 2019 to close all coal-fired power stations by 2030, although it abandoned a cap on production last month until 2024 to address possible Russian natural gas shortages. Its government has also pledged at least €35 billion ($52 billion) for renewable energy and energy-efficiency measures, though it has been criticised by environment groups who say the emissions cuts ordered by the court were not delivered in time. Urgenda supports lawyers and activists around the world in similar litigation, and the London School of Economics has counted more than 73 related 'framework' cases challenging governmental responses.

When I ask Best how it feels to be involved in all this, and leading the Pabai case, he smiles: 'This sort of case is the reason I studied law.'

The Pabai matter is not the only climate-linked action afoot in the Torres Strait. In the quiet beer garden in a hotel on Thursday Island, I meet Yessie Mosby, a member of the Kulkagal nation and traditional owner from Masig Island. Mosby and his 11-year-old son Genia are on their way home from Sydney, where he opened the First Nations show at Australian Fashion Week, blowing a conch shell – a call to attention or action in many Pacific cultures – while dressed in a *dhari* warrior headdress made from cassowary feathers.

The 39-year-old is one of eight traditional owners, known as the Torres Strait Eight, who in 2019 took the federal government to the UN Human Rights Committee for failing to protect the islands from climate damage. A decision on the complaint may be announced this year.

There have been many changes to tiny Masig and the surrounding ocean since Mosby was a child. The land and sky have long told Kulkagal what to do: when a certain bird flies to Papua, it's time to backburn and prepare the gardens; when a particular insect shrills, the rain will arrive within days. But now the mating season for *waru* (turtle) has shifted, the water is warmer, and the annual rhythms of animals and plants are disturbed. And, as on Saibai, when the wind combines with a high tide, the sea washes into the cemetery. 'I was walking with my children and picking up our ancestors' remains off the beach, like shells, when I knew something had to be done as soon as possible,' he says.

On the drive back to Saibai township from the cemetery, Pabai and Kabai say how much they appreciate the support they've received since launching their case, from both their community and the Australian public. 'Whenever I get stressed out I talk to the elders, and they say, "We're going to do this together",' says Kabai. 'I don't keep everything to myself. It's better to share your thoughts, your views, with others and they can direct you.'

Pabai takes a big picture view. 'If we don't have support from the organisations and the people out there, we lose everything. If we don't win in this case, that means our islands will be underwater, and no one will be listening to what is here in the Torres Strait. It's not only for us, it's for everyone.'

✱ *A city of islands*, p. **60**
A subantarctic sentinel, p. **85**

LONG COVID: AFTER-EFFECT HITS UP TO 400 000 AUSTRALIANS

Bianca Nogrady

Today, James is feeling about a two out of five. The young lawyer is reasonably articulate on the phone. He might be able to read a single news article today, maybe even go for a short walk. On a 'five out of five' day, he can do four hours of lighter duties – analysing legislation, some emails – but nothing like the intense legal work he was doing fulltime eight months ago.

When he's at zero, he's immobilised on the sofa, unable to compose a two-sentence text message or string together enough words to speak to his partner, Matilda. Getting up to go to the bathroom is a marathon. The sound of a plastic bag crinkling is like someone screaming in his face.

James, whose name has been changed, spends half the day battling to stay awake. At night, he can't sleep for more than four or five hours. His heart is constantly racing and he is plagued by intense headaches. If he has two good days in a row and makes the most of that, the price is being back at zero for the following three or four days: 'Basically a living, breathing shell,' he says.

This is long Covid. James's symptoms are severe, but an estimated 400 000 Australians are in a similar boat. With Omicron surging again, that number is going to increase. Estimates of the rate of long Covid in people who have been infected with SARS-CoV-2 range from 5 per cent all the way to 50 per cent, depending on the definition, the population studied, and the time frame used.

'I think we land in the area of, crudely, let's say 10 per cent of people who've been infected having long Covid at the three-month point, which is the critical point where it's really "long Covid",' says health economist Professor Martin Hensher, the Henry Baldwin professorial research fellow in health system sustainability at the University of Tasmania.

More than 8 million cases of Covid-19 have been diagnosed in Australia since the start of the pandemic and, says Hensher, 'the fundamental problem is 10 per cent of a big number is a big number.'

Hensher and colleagues did some modelling in the second half of 2021 – before Omicron appeared – looking at the overall burden of Covid-19 in Australia in terms of disability-adjusted life years, or DALYs, where one DALY represents the loss of one year of good health. Their modelling included DALYs lost because of death, severe illness, mild illness and ongoing illness.

They found that persistent illness, in the form of long Covid as well as conditions precipitated by Covid-19 infection, including diabetes and heart disease, was likely to account for about half of the total health burden of Covid-19. Long Covid alone represented about 10 per cent of Covid-19's health impact. That was before Omicron.

Now, Hensher thinks, long Covid will make up an even greater proportion of the health burden of Covid-19, because deaths are going down but infection numbers are going up.

He and his colleagues are gearing up to rerun their model with more recent data, but a key challenge is defining long Covid. He says 'the critical missing piece in all the current discussion' is clarity around how many people are really disabled by it versus those people for whom it is 'thoroughly unpleasant and highly inconvenient'.

In December 2021, the World Health Organization put out its first clinical definition of long Covid, or what it calls 'post Covid-19 condition'. Key features are probable or confirmed SARS-CoV-2 infection, and symptoms that have lasted for at least two months and aren't explained by any other diagnosis. Those symptoms might have persisted after the first, acute phase of infection, or started just after recovering from the initial infection.

At Anthony Byrne's long Covid clinic at St Vincent's Hospital in Sydney, they treat patients who have had persistent symptoms for at least 12 weeks, 'but the key thing for that is that they're symptoms that weren't there before, and they're there now, and they're not explained by something else', says Byrne, a chest physician and associate professor at UNSW Sydney.

Byrne has treated patients from age 18 to 80, men and women, in all states of health or illness before Covid-19 hit them. 'There is a broad spectrum of what a long Covid patient looks like,' he says. 'It can look like anyone.'

What they do have in common are symptoms of crushing fatigue and lethargy, breathlessness and 'brain fog'. Some also have chronic pain, headaches or insomnia. These symptoms have to be pretty severe: James qualified as a patient, but his partner, Matilda, did not, despite months of being unable to work more than a few hours a day, drive for more than 20 minutes without pulling over to rest, or walk and text without needing to sit down. Both Matilda and James caught Covid-19 in early December 2021, so they're now past the seven-month mark of persistent, debilitating symptoms.

The symptoms of long Covid will be unpleasantly familiar to anyone who has experienced post-viral fatigue or chronic fatigue syndrome/myalgic encephalomyelitis. There's a number of important similarities between all these conditions, says Dr Natalie Eaton-Fitch, a research fellow with the National Centre for Neuroimmunology and Emerging Diseases at Griffith University. 'Both ME/CFS patients and long Covid patients present with brain fog or the cognitive difficulties,' she says. 'There's the post-exertional fatigue, the respiratory symptoms – they're also present in ME/CFS, as well as long Covid.'

ME/CFS has been around a lot longer than long Covid but

researchers are still largely at the hypothesis-generating stage of explaining its underlying causes. However, given the massive impact and cost of long Covid, and the substantial global research effort to understand and treat it, the hope is that scientific explanations will be found much more quickly.

There are some leading scientific hypotheses about the underlying causes of long Covid that fit broadly into the categories of immunological or neurological explanations.

The first thing to note about long Covid – as well as post-viral fatigue – is that despite the crippling symptoms, people with the condition have relatively normal function in all the organs that seem to be affected. It's a paradox that has puzzled infectious diseases physician Professor Andrew Lloyd for 30 years of researching and treating post-viral fatigue.

'People report a lot of difficulty, but if you try and measure it – this is true in post-viral fatigue as well – you can show a little bit, like milliseconds, of delayed reaction time or slightly reduced performance in ... what we'd call executive tasks, but they're pretty subtle,' says Lloyd, head of the viral immunology systems program at the Kirby Institute, and director of the UNSW Sydney Fatigue Clinic and research program.

Similarly with breathlessness: a person with breathlessness related to asthma or lung disease will be puffing hard after going up a flight of stairs. But a person with long Covid can walk up that same flight of stairs and, although they feel subjectively breathless at the top, they can still have a conversation.

'If you do brain scans, and structural scans like MRI, there's a brain there and it looks for all intents and purposes normal in size and shape,' Lloyd says. 'There's not much to find structurally, which leads us to suspect that it's a functional disturbance.'

Lloyd points to post-viral fatigue research using functional magnetic resonance imaging, which gives an indication of levels of activation in different parts of the brain in real time while someone performs certain tasks. This suggests that people with post-viral fatigue have a greater degree of activation during tasks, such as tapping their finger or thinking about something specific, compared with people without post-viral fatigue. 'It's like the system's not operating as efficiently as you'd like to get the output,' he says, 'and so you've got to work harder.'

So what's causing this apparent inefficiency? One theory is that inflammation in the brain is the culprit. Neuroscientist Dr Leah Beauchamp, from the Florey Institute of Neuroscience and Mental Health in Melbourne, first became interested in the neurological effects of Covid-19 when she started hearing reports of people losing their sense of taste and smell. As a researcher in Parkinson's disease, she knew those losses were also a possible harbinger of neurodegenerative disease, and that worried her.

The initial loss of taste and smell that characterised acute SARS-CoV-2 infection in the pre-Omicron era was considered par for the course of an infection that got into the cells lining the nasal passages. 'During other viruses, you do lose your sense of smell but again, it's acutely: it's for a couple of weeks or a month, because you're really inflamed and congested,' Beauchamp says. The persistent loss in long Covid isn't normal, and suggests that inflammation has actually penetrated into key regions of the brain associated with olfaction. 'These factors that are inflammatory from your system can get into your brain and it triggers a cascade,' she says, 'and once you've triggered neuroinflammation, it's very hard to rein that back in.'

The other support for a critical role of inflammation comes from immunology. At the Kirby Institute, Dr Chansavath Phetsouphanh and colleagues were comparing immunological markers in people with long Covid against those of people who had experienced Covid-19 infection but had recovered without persistent symptoms. This showed elevated levels of certain proteins involved in the immune response to viral infections, called interferons, even at eight months after their initial infection, which suggests ongoing inflammation.

Patients with long Covid were also missing a type of immune cell called a true naive T cell, and the researchers suspected these were being constantly switched over into an activated T cell. 'They're not meant to be activated,' Phetsouphanh says. 'It just goes into what we observed with interferon, so it seems like there's ongoing inflammation in long Covid patients, and it's activating the immune system.'

Inflammation is a modern bugbear, seemingly behind a host of chronic conditions including diabetes and heart disease. While drugs known as anti-inflammatories exist, they are hopelessly outgunned by system-wide, entrenched inflammation. This makes treating and managing long Covid difficult but not impossible.

'We know that over the last two-and-a-half years, people get better. They do get better. But it's everything in super-slow motion,' Byrne says. 'So what we're looking to do is reassure, diagnose and assist with getting people back.'

For James, that assistance has taken the form of breathing exercises, tai chi, cold therapy such as ice baths, stretches, and being very, very careful with how much he tries to do in a day. 'A lot of it is finding the right balance and giving me a framework to figure out – if I'm trying to return to work, which I am, do I try to do this much exercise and this many hours, or fewer hours and more exercise, or what kind of exercise?' he says.

Despite countless blood tests, scans, and visits to neurologists, cardiologists, physiotherapists, psychologists and sleep physicians, no obvious treatment targets have yet presented themselves. The most emphatic point made by many experts is that the best treatment for long Covid is prevention – not getting Covid-19 in the first place. Byrne has observed that the more acute Covid-19 symptoms someone gets, the more likely they are to go on to long Covid, which makes a strong case for avoiding infection or getting vaccinated to reduce the risk of severe infection.

Another clarion call from the medical and scientific community is for us to not repeat the mistakes of the past when it comes to recognising and managing long Covid. 'We have some terrible examples from the way that we've responded to prior issues like chronic fatigue syndrome, where people have not acknowledged the physical and the psychological impacts of that,' says Professor Maree Teesson, director of the Matilda Centre at the University of Sydney.

Given the massive impact long Covid has on mental health, such as brain fog, anxiety, depression and insomnia, Teesson hopes that patients with the condition won't be subjected to the same stigma that has met patients with ME/CFS, and that they will be given all the best, evidence-based treatments available. 'I'm really hopeful that we've got a government and we've got a Health minister who is already discussing long Covid and is hopefully prepared to respond.'

She also notes that Australia's already strained health system, especially mental health services, will not cope with a huge influx of chronic disease. 'Quite frankly, our mental health system is not coping at the moment, let alone coping with another 400 000 people with additional mental health symptoms associated with long Covid.'

Professor Andrew Baillie, convenor of the Long-Covid Australia Collaboration, makes a similar point. 'The NDIS isn't really, I don't think, ready for this.'

Baillie, who is also a clinical psychologist at the University of Sydney, would like to see income support provided for people with long Covid who have used up their sick leave but are still unable to work.

He also says Australia needs to know exactly what it's dealing with. 'I think at the moment somehow we've got our head in the sand; we're not even looking, so we don't know how big a problem it is.'

There are currently no registers tracking long Covid cases, or any centralised data collection at the state or federal level. In contrast, Britain has implemented large-scale household surveys that are giving some sense of how widespread the condition is.

Most importantly, however, Australia needs a plan for long Covid that is agile and responsive. 'Covid is not going anywhere,' Baillie says, 'and the more Covid cases we have, the more long Covid we're going to have.'

✱ *A whole body mystery*, p. **113**
   Bats live with dozens of nasty viruses – can studying
   them help stop pandemics?, p. **271**

NOISELESS MESSENGERS

Rebecca Giggs

The moths, when they came, were said to appear first like sea fog massing above the ocean. Lighthouse keepers along the south-eastern edge of Australia warned of beacons so darkly swarmed that navigators doubted their bearings. Ferryboats were burdened by thousands, wings ablur. Some moths hung in clusters off the precipitous coastal cliffs, living icicles, dripping with more moths. The moths, as a myriad, moved in: at nightfall they 'swept over the suburbs in clouds', wrote one reporter for a Sydney tabloid, *The Sun*. Descending into tea trees and turpentine gums in Gosford, the seething of the moths gave the impression of bough-shaking winds when all else was motionless.

If the moths' light-seeking caused disruption in the darkness, their urge to seek shelter when the dawn broke made them a more invidious presence yet. So many of their oily bodies were crushed on train tracks that slowdowns were mandated to stop locomotives slipping from the rails. They jammed the circuitry of elevators, spoiled gatherings. At a Government House garden party in Canberra every cake was seen to be decorated with moths. The moths entered people's houses. They crept behind upright pianos, into radio sets, betwixt the slats of venetian blinds. They got between the mattress and bedsheets, and huddled in the pockets of dress suits. In kitchens, gutted fish were found to have bellyfuls of moths. One year, churchgoers counted 80 000 moths on the windows of Saint Thomas's prayerhouse in North Sydney. Services were cancelled for seven days, the building sealed while the moths congregated under the eaves. People reached for words like *visitation, marvel.* The less-enraptured said: *plague.*

These were migratory moths, called bogong moths, and through the early 20th century few people could say with confidence where they came from. 'Noiseless messengers', the *Argus* newspaper deemed the bogongs in 1916 – '*noiseless messengers* sent forth to flicker ghost-like through space, and collect the news of other worlds.' Truth was, the moths had their origins underground. After frail frosts, and when the early spring was wet, great throngs of moths emerged from pupae in the soils of lowland southern Queensland, and in western and north-western New South Wales. Stirred by some ephemeral cue (temperature, day length, barometric pressure), the moths took off. Though no single individual in their generation had ever made the journey – and while each moth's brain is scarcely a speck – the moths set out to travel over 1000 kilometres by instinct.

Some years were sparser than others, but when conditions favoured the moths, there could be over 4 billion on the wing. Passing through the railyards of Newcastle, they obscured electric signals. As far south as Mirboo North in Gippsland, men complained of needing to move agglomerations of moths off the paddocks by the shovelful. Lacking the necessary mouthparts to chew leaves, they did not skeletonise plant life as locusts do. Instead, the moths relied on floral sugars to power them, supping thin streams of nectar via their proboscises, along with lerp – a type of honeydew extruded by psyllid insects. Each moth could only ingest a skerrick of sweetness, but they were so numerous that apiarists nonetheless found they had to sustain their bees on syrup after the moth front had passed by, taking with it much of the nourishment otherwise found in flowering yellow box, red box, grevillea.

Many moths were killed – by nightjars and frogmouths, by high winds, by sizzling up in light fixtures, and by slapping hands – but there always seemed to be more to come. They were impervious to knock-down sprays. Any attempt at sweeping them from a surface left behind black marks. In Dubbo in 1919: the moths 'destroy[ed] the happiness of many a domestic circle, and by their dying help[ed] to increase the cost of living'. Removing moths from the home was nearly impossible. One might as soon have tried to net a mist and tow it back out to sea.

Yet, not that long after they arrived, the moths disappeared from the cities, like a nightmare dissipated on waking. Where had the moths gone to? From at least the end of the last Ice Age, the moths have taken their leave, every year, to go into hiding in the Australian Alps. The bogongs chase the cold. From several reservoirs inland of the Dividing Range, they funnel together to seek refuge from the hottest weather by climbing up above the tree line into chilled crevices and grottoes in the high-altitude scree of the Snowy Mountains, the Victorian Alps and the Brindabellas. When they finally enter their encampments in the granite and basalt, the moths settle on the rock in a tessellating pattern. Each moth, a jigsaw piece, tucks its head under the hindwings of the one in front of it, until there is a wide brocade of moths that can extend for maybe 80 metres squared, or more. If they blanket the interior of a cave, it can come to feel like a softly padded cell. The moths enter a torpor called 'aestivation'. Aestivation is the opposite of hibernation: it is done to circumvent the swelter, not the snow. The moths are mostly motionless. Intermittently they jiggle their wings. A handful might take a turn in the open air each night before settling into their long tranquillity again. Bogongs live much longer than the average moth: between eight and nine months. They will stay across the turn of the new year, before returning to breed, lay eggs, and die where they were born, in the cracking clays far away.

The Jaithmathang, Gunaikurnai and Taungurung peoples knew of this migration – and had known since long before European invasion. The moths' name, 'bogong', comes from the southeastern Indigenous language groups. *Bungung* denotes a moth of the mountains, or the mountains of the moths, and the brown color that envelops both. The aestivation of the moths drew Indigenous Australians into the lowlands and foothills of the Alps, land that was cyclically inhabited by the Traditional Owners who have been its continuous custodians, and in its care, for all of the time-span the Dreaming encloses. At several waypoints along the moths' passage, people stupefied the insects with smoke and cooked them in fires before grinding their bodies into a paste and fashioning long-lasting patties. For the moth hunters, bogongs were a seasonal cornerstone of their diet. The custom was so widespread that it changed the appearance of the landscape: the ground was rumpled like a quilt from where the firepits were dug year after year. Significant law, intergroup consultation and ceremony are associated with the occupation of the high plains at moth-harvest time. Dispossession and colonial violence disturbed these practices.

For centuries the bogong moths streamed back into the caves, slept, and vanished again when the weather cooled. The moth hunters illustrated facets of the feast on rock walls. Later came scientists, from research institutions and universities, to study the moths across the tors of several summits. The scientists noted that the moths smelled sweet like molasses when they arrived, and thereafter awful, like compost. Then, in the summer of 2017/18, the bogong moths, prolific as they had been for all the years prior, vanished more completely. Where once there had been hundreds of thousands of the insects – a juggernaut, a moving nimbus – now the night air stood empty. On Mount Morgan and Mount Gingera: no moths. A cave long favoured by moths on a boulder outcrop near South Ramshead, in the Kosciuszko Main Range, saw only a smattering very deep in the far reaches; likewise in known habitat on Mount Buffalo. The next summer a single live moth was found on Mount Morgan. Three moths made it to Mount Gingera. With a note of terror the scientists reported that within a few short years bogong numbers had declined by 99.5 per cent. The International Union for the Conservation of Nature (IUCN) formally listed the bogong moth as endangered. What has happened to the moths, and what will happen in the mountains if they are not restored, has much to tell about how we envision the manifold crises we are connected to, and the scale on which they occur.

I had begun reading about the moths at a time when we were compelled to stay indoors, to wait out a state of semi-dormancy in lockdown. It started with a dream, the dream everyone seemed to be having that spring, of a billowing swarm of insects. In mine, the insects were seen from far off – a murmuration knotting and unknotting on the horizon, banking into mountainous peaks that shivered and collapsed. At a distance I registered only curiosity towards them, but then the insects started to collect together, to concentrate, and cascade down upon the buildings of the city into the streets. A horrible fear gripped me. Long minutes after waking I could still feel the prickling of legs and wings landing all over my skin.

Researchers who study dreams have observed that, since the outset of the pandemic, the imagery of insects has proliferated in sleeping minds. A virus is not a living organism, but we sometimes call it a 'bug', so the theory holds that bugs (of the arthropod kind) were a ready-made metaphor to visualise an invisible threat. It was easy to imagine an insect horde passing through people's heads on their pillows at night, like a storm traversing the eastern seaboard. Friends detailed variations on the theme: one in which ants overtook a classroom, one where a train carriage filled with hornets. My aunt dreamt that she was compelled to hold a single insect in her mouth as she moved through a crowd. *Bitter*, she said.

Insects are bonded to ideas of mortality the world over, being both decomposers and natural transformers: scarab beetles, a feature of funerary art in Ancient Egypt. Jade cicadas, placed on the tongues of the deceased in the Han Dynasty to ease the transition into the afterlife. In the portraiture of pre-17th-century Europe, the addition of a fly signified the subject was no longer alive. To be touched by insects – to be traipsed by a lacewing, or to cup a centipede in your palm – is a morbid sensation, I think; a foretoken of the moment the body ceases to sense the lightest contact, before it begins to turn, in time, to stillness, to ash on the wind, or dust on the touchpaper texture of a moth.

After the curfew each evening, I sat at home with the lamps on, watching whatever pinwheeled and buzzed against the glass. The borders were closed and a 5-kilometre travel limit had been imposed, in addition to social distancing measures. Life seemed to have contracted down to very little. The window had become my public square, I thought. Could I get interested in what was there, within arm's reach but out of touch; what visited the house, and where it came from? Mostly, it was moths. Between me and the moon, moth after moth. Unearthly.

What I then knew about moths wouldn't run to more than a sentence or two. Night insects, yes, the idiom 'like a moth to a flame', metamorphosis – full stop. I found a guidebook with pictures online and scrolled through it, surprised to discover that what distinguishes moths from butterflies is not, as it transpires, their circadian habit. There are many day-flying moths. Dark-loving butterfly species populate the understories of rainforests in the country's north. Neither is the difference a question of drabness. Moths can be dotted with vivid iridescence, as if they'd dragged their wingtips through petrol: there are green, blue, violet, pink, marigold and piebald moths. One is the bright orange of a traffic cone; another is banana-yellow with blood-red eyeballs. In dense, low-lit woodlands, a few moth species have evolved to be almost completely transparent, a form of camouflage that means you don't so much see the insect as notice a ripple crossing the leaf litter.

No, finer details divide moths from butterflies. Both belong to the insect order Lepidoptera, but as a general rule butterflies tend to have club-shaped antennae, where those of moths are more thready, or look like wincey bottlebrushes: an attribute that helps males pick up wafting pheromones during the breeding season.

Even if they have narrow wings, moths are also more liable to have bristles on the surface between fore- and hindwing – a kind of Velcro that keeps all four wings aligned in motion. And, with exceptions, butterflies are inclined to rest holding their wings sandwiched together vertically, whereas moths idle with their wings folded over their backs, like a collapsed tent, or held out flat as per a Rorschach blot.

In all its ingenuity, evolution devised a single organism capable of living two lives, at two speeds. First, the reclusive homebody, the caterpillar, a fleshy little digester in a vast empire of leaves, reliant on a plentiful if low-nutrient diet. Second, the winged moth. Extremely mobile but slight and soon to die, moths either eat nothing during their maturity or are dependent on high-energy but scarcer foodstuffs, such as sap or nectar. As a strategy, this duet of bodies has proved so successful that insects exploiting it have been around since the middle–late Jurassic. Specimens of the most primitive moths, the Micropterigidae (nine species of which live on in Australia), have been found clenched in amber from a time near to 200 million years ago, when they might have been fodder for flying dinosaurs. With such a deep evolutionary past to pull upon, Australian moths have diversified into a plethora of specialists.

Some are marbled, some are woolly. Some look like pieces of rotting wood, bird droppings or thorns. A moth that appears to be a splotch of turquoise mould reveals startling coral-coloured hindwings when it flies. Another trails streamers that baffle birds chasing it through the air. Here is a thorax as purple and shiny as plum skin; further on in the guidebook, a moth with a shaggy, bear-like countenance. Some roll their wings up to look like tiny antlers. One seems pixelated, like a rasterised object in a video game. Australia is home to between 20 000 and 30 000 moth species – almost as many as there are flowering plants – but only some 400 butterflies (a 'depauperate insect fauna').

Some moths can only be told apart by their gait when walking, having either a 'waddling' gait, a 'dancing' gait, or moving 'quick-slow-quick' as in a foxtrot. A mothtrot. A few engage in tactical mimicry: of wasps, of repellent beetles, of less-edible moths that are their cousins. Some are furnished with hairs capable of triggering allergies and anaphylactic shock. Though the preponderance are herbivorous as larvae, there are also carnivores and frugivores. One moth tricks meat ants into carrying its caterpillars into their nests, where the larvae dine delicately on infant ants.

Among this spellbinding Australian bestiary are some of the world's largest and heaviest moths. *Coscinocera hercules*, the Hercules moth, is found in northern Queensland and can grow to have a wingspan of 36 centimetres – the diameter of a car's steering wheel. Caterpillars of the Hercules moth feed in bleeding heart trees, and then pupate for two years. The adult moth, which moves somewhat floppily, like a sunhat, lives only two days. Earlier this year construction workers sinking the foundations for a school in Mount Cotton disturbed a giant wood moth, *Endoxyla cinereus*, the heftiest species yet identified by science and not uncommon, though it is rarely seen. A builder balanced it on the tip of a saw for a photograph – a moth the size of a catcher's mitt, its dusky legs dangling.

And as for bogongs? I remembered only that they were famously innumerable and transient. Having grown up on the west coast, I had never seen one. Now I wanted to. I came to the entry for 'bogong moths' in the guidebook. *Agrotis infusa*: the moth's Latinate name evokes 'infused fields', a head nod to the fact that bogongs pupate mostly in croplands, in chrysalises that are sepia and translucent, like varnish on a coffin. It was the right time of year to see one. Teak brown with a fuzzy sort of cape extending over the back of its head and collar. Bogongs are small, with a wingspan of about 5 centimetres, and they have reflective eyes – a feature that characterises members of the Noctuidae family, the moths Americans call 'owlets' because their gleaming eyes bring to mind those of owls swept by flashlight. Reading on, I learn that male bogongs have antennae that resemble hair-combs—a hallmark that is only observable with the aid of a magnifying glass. Otherwise moths of this kind are unexceptional. Bogong moths are easy to miss, easy to mistake, save for this feature: on each wing are two pale dots, one slightly elongated like a comma. A moth adorned with semicolons.

It was the semicolons that set me off in the end: a gesture to the branching nature of sentences, and therefore of time; the possibility of subclauses running into the future, paths taken and not taken. The idea nestled into me. It was pleasant to think of something so small as a bogong moving out there, from state to state, when all else was grounded. More gratifying yet was the picture that came to me next of a stranger, their gaze alighting on a bogong moth a long way away. That person becoming verily engrossed, following the moth's mid-air helixing until it spiralled off into the dark, and then, in time, *that same moth* appearing to me, conveying the tiniest of contact-highs; the vision of someone elsewhere, grown watchful of insect life. Maybe there was another woman I had never met, a woman who sat by her window, even now, looking for something to wander across her reflection and bring her back to life. Perhaps she pressed her thumb against the glass and felt the faint vibrations of a moth squaring on the other side. If she thought of someone like me then, with her thumbprint fading, I hoped it made her feel she was not alone.

What was it that Virginia Woolf wrote, in her famous essay addressing a moth flown to her window ledge? 'Just as life had been strange a few minutes before, so death was now as strange.'

On the natural exodus and ingress of insects, science was, for much of history, limited to guesswork. Invertebrates have proved hard to track for all the obvious reasons – tininess, dispersion, difficulty telling kindred species apart – and, too, because many insects shapeshift across their lifecycle. Especially elusive have been 'noctivagant' or night-wandering insects, those that take advantage of a drop in thermal currents after sunset to wend their way by cover of darkness. Even in very large numbers, nocturnal insects can pass by unnoticed. One morning the bugs are just *here*: An old magic is frisking the shrubbery. No wonder the thinkers of antiquity held that many such insects were inert matter sparked to life. Aristotle and Pliny both suspected insects of being spawned by 'spontaneous generation' from origins in fire, mud, dew, snow and sand. Cicadas were fancied to issue from the spittle of cuckoos. A felicitous wind, most of all, was understood not just to transport scores of insects, but to fabricate them out of thin air.

The fact of insect migration – not to mention the basic biology of metamorphosis – has since been well substantiated, but though researchers have known for over a century that insects undertake long-distance travel, the misguided belief that such movement is entirely passive and dictated by the winds persisted into the 1980s. Some insects *are* at the mercy of the weather: *Persectania ewingii*, for example – called the 'armyworm' before it transmogrifies into a silken, buff-coloured moth. *P. ewingii* is tossed by spring winds from South Australia across the Bass Strait, only to be saved from perishing in the ocean by the appearance of Tasmania (or the tussock-covered shores of Macquarie Island). But as the technology capable of monitoring insects has improved, it has become clear that several winged insects – including bogongs – sense, and selectively choose, which air currents to ride, some forming massive, multispecies 'bioflows' at high altitude. Radar has given entomologists a better picture of the little life gliding by far above us. What they've seen up there is, frankly, astounding. The spectacle of animal migration may be typified by herds sweeping over the Serengeti, but most terrestrial migrants are insects – by number of individuals and, perhaps more surprisingly, by mass. One study showed that each year in south-central Britain 2–5 trillion high-flying insects migrate over an area roughly the size of Tasmania. Together, those insects have an estimated biomass greater than that of the nation's migratory songbirds.

Indeed, the volume of insects up in the air is so tremendous that researchers have suggested thinking of them as 'the plankton of the sky': a constant particulate, bobbing overhead.

At the same time as our means of quantifying aerial insects has been upgraded, so too has our understanding of the impact of insect migration down on the ground. Large, travelling vertebrates, including elephants, caribou and wildebeest, have long been known to link up ecosystems, transporting energy and nutrients (as well as pathogens) across the terrains they traverse via routine grazing, defecating and dying. In recent years it has become clear that smaller animals, in a large enough profusion, can likewise leave a lasting impression on the landscapes they pass through.

Though the movements of insects are often more covert than those of large mammals or fish-runs, their transit can shape an ecosystem in durable ways. Tiny beings have system-wide effects over generations. Because many winged insects are pollinators, they create gene flows between plants they alight on along the route of their journey. In Spain, for instance, endangered violets surviving in distant islands of habitat are genetically enchained together by the migration of hummingbird hawkmoths. Each hawkmoth threads flower to flower. Trees along a 250-kilometre stretch of the Ugab River, in Namibia, have genetic linkages that flow with the easterly movement of fig wasps. Billions of pollen grains are shifted southward each year by high-flying hoverflies in the UK, some lifted over the English Channel to the landing-pads of flowers in Europe – a targeted haulage the wind alone could never achieve. Insect migration also acts as a resource pulse in environments where endemic vertebrates feed off migrants. An influx of migrants – flies, beetles, butterflies, locusts – can intermittently decouple predator–prey dynamics, permitting prey that is low on the food chain to rebound, when those that eat them discover a more effortless meal of new arrivals.

Many migrating animals trail their favoured weather conditions, though the bogongs must be one of only a handful of species to zero in on microclimates as pinched as those in the stony nooks they seek. To trigger aestivation, the moths need to find a protected place with a round-the-clock temperature below 16 degrees Celsius; a chill that must remain constant for months. In a way, the moths' journey is not to the mountains, but to cupboards of air from a former geological age, very cold air which is found during the summer now only at high elevations.

How a bogong knows where to go – how a single moth intuits the existence of a microclimate hidden in the rocks hundreds of kilometres away, far above sea level – is a perennial mystery. It cannot have been taught by its parents the way foraging bees relay travel routes via dance: a bogong's forebears offer no nurturance or guidance, having died before the infant moth hatches into a caterpillar. If the homing instincts of swallows and sea turtles are astonishing, consider that a bogong moth makes its journey as part of a single generation, without elders, in complete naivety. Scientists have learned that the moths can apprehend the planet's magnetic fields and that they take their orientation from stellar cues, skills that are rare, if not unique to the moth alone, in the insect kingdom. Sometimes they fly *against* the wind to maintain their bearings, a determination that is especially startling when you consider how small a moth is, and how powerful the wind.

When the bogongs arrive in the Alps, mountain-living mammals rush out to gorge on moth flesh, some with young in the pouch. Broad-toothed rats dart about. So too *Antechinus* (a marsupial shrew) and the rare pygmy mountain possum. Birds binge, ravens especially. Feral pigs have developed a taste for moths. With evident cunning, the pigs remember the richest places to find them and wait there patiently for the moths to arrive, their clammy noses trembling. Nocturnal as the moths may be, the surge of bogongs up the mountainsides has been calculated to amount to an injection of energy into Australia's alpine ecosystem that is second only to sunshine.

If the bogong moths were once called a *plague*, it was not because they brought sickness. Rather, their influx frightened city people with its obtrusive, frenzied persistence. A world of our design promised to restrict insect life to the margins; but having been besieged by moths, one could be forgiven for believing, however fleetingly, that our houses were less of a permanent fixture than the cascades of insects that opportunistically sought shelter within them. That we might have been living in the moths' home all along proved an uncomfortable notion; one that implied we had less right to control what was excluded, and what was invited in, than we might have supposed. Whose world was this? Who belonged to it – and who only *claimed* the world belonged to them? Ideas like these were perhaps what the Belgian playwright (and armchair entomologist) Maurice Maeterlinck had in mind when, over a century ago, he wrote:

> The insect brings with him something that does not seem
> to belong to the customs, the morale, the psychology of our
> globe. One would say that it comes from another planet,
> more monstrous, more dynamic, more insensate, more
> atrocious, more infernal than ours.

But then, Maeterlinck didn't live to see that the planet 'more monstrous, more dynamic ... more atrocious, [and] more infernal' could, in fact, be our own Earth transformed. In the climate era, it is the disappearance of insects, not their efflorescence, that haunts us most of all. A global research review completed in 2019 found that 40 per cent of known insect species are declining, and a third are endangered. Countless insects have suffered a reduction in range as well as numerousness – which is to say that, where they live now, many insect species also live less densely than before. In his elegiac memoir, *The Moth Snowstorm* (2015), British author Michael McCarthy coined the evocative phrase 'the great thinning' to capture this overall loss of bug biomass. Once-common minifauna register today as exotic, hard to spot. People are more likely to assume endangered insects are interlopers, invasives, both because it is our habit to overrate our familiarity with the nature under our noses, and because those creatures that surprise us *are*, with increasing frequency, newcomers and novelties. Among those insects that are best able to pursue conditions beyond the boundaries of their former territory – driven along corridors of climatic change – some will be evolved by the journey so that they end up not just 'out of place' but in altered bodies. (This is true of the speckled wood butterfly in Scotland, which now displays larger wing muscles along the northward edge of its habitat, where warming is drawing the butterfly into longer flights.)

Of all insect vanishings, that of the bogong moth is remarkably, terrifyingly precipitous. Insects more typically die off where habitat disappears. Those species that are most at risk are specialists with a circumscribed range and ordinarily small populations that rely on limited food plants, or are allied to threatened organisms (the rhinoceros stomach botfly can only thrive so long as there are ample rhinoceros undercarriages to parasitise). Colony insects – found in hives and anthills – are additionally vulnerable to the spread of pernicious diseases and mites because they live in such close and constant proximity to one another. As to moths particularly, traits shared by those taxa that have suffered the most rapid declines are: large wingspan, low dispersal ability, short flight seasons, and a genetic phenotype that has become 'canalised' (that is, narrowed), with fewer of the subtle perturbations seen in species that occupy multiple geographic sites.

By all these measures, the bogong moth should be secure. Insects with wings are better able to leave a hostile environment. A few can enter into a time capsule, as those insects that burrow deep into the woody tissues of trees to pupate can sometimes wait out the barren period after a bushfire. Insects that have historically maintained very big populations have long been assumed to be comparatively insulated from extinction. Indeed, monitoring plentiful insect species not only presented technical challenges in the past but was also considered to be of low importance because models of decline held that organisms vanished at an observable rate, moving from abundant to sparse ('thinning'), and from secure to threatened, with interventions possible at several stages along the drop. What has happened to the bogong moth challenges that model of extinction and suggests that even very common insects can be at risk.

The moth's habitat has not disappeared, taking the moth along with it. The mountains loom still; the plants the cutworms eat continue to grow in the fields (bogong larvae consume wild capeweed and crops, including cauliflower, silver beet, lucerne, flax and other cereals). Over the decades agriculturalists have attempted to stamp out the larvae: with hellebore and pyrethrum, then with arsenic mixed into bran, or with a pound of DDT per acre, and by running stock and passing heavy rollers over the earth. The moths survived it all. In the last 40 years, the southward expansion of rice and cotton growing in the bogongs' natal grounds along the Murrumbidgee River, and in the Murray Valley, has no doubt curtailed their range (bogongs do not eat cotton, and the moths cannot pupate in flooded rice fields), but scientists estimate the impact of that agroeconomic shift resulted in a cull of around one-sixteenth of the population. A blow, but far from a fatal one.

What has changed, what is changing, is that the topographic features of the landscape, coded into the bogongs' migratory instincts, are decoupling from the climatic conditions necessary for the moth to complete its life-cycle. This is true at both ends of the journey. The emergence of a butterfly or moth from its pupa is called 'eclosion', and from 2017 to 2019 (and in some locations, for far longer) many of the areas where bogongs eclose were stricken by extreme drought. Temperatures climbed to highs with precedents seen only in models of paleoclimatic weather 2–3 million years ago. The Federation Drought, the World War II Drought, even the Millennium Drought, all paled by comparison. In earth robbed of moisture, moth pupae were surely extirpated by dehydration. Larvae had less to eat when crops failed, and those robust plants that did endure were guarded more preciously by agricultural workers. The diminished number of adult moths that departed on their migration would have returned to lay eggs under those same conditions, on fields ablaze with heat.

In the mountains, too, climate change is driving an uptick in temperature. Seasonal snowmelt in the Australian Alps has inched forward almost three calendar days per decade from a baseline set in the 1950s. Mountain-living animals such as *Antechinus* – which breeds beneath snow cover in winter and emerges famished – now confront a metabolic rift. As the snow inches back earlier each year, surfacing females encumbered with young discover that there is little to be gleaned just when they are most in need of calories. The gap may end up being too long to bridge, when what moths there are remain on the wing kilometres away. So the changing weather is a force of disunity: the moth world and that of its predators breaks apart. Any life on the mountain that relies on the moths is embrittled.

Raised elevations are always chillier than the foothills and plains, and so the effect of climate change on the mountains is to draw cold-adapted habitat upwards. Some plants and animals found previously in lower belts of country creep upslope towards higher altitudes, often moving into more crowded terrain (as the surface area of a peak narrows into a summit). Organisms found at the tops of mountains face the evaporation of their domain into midair when apex conditions exceed their tolerance. For the bogong moths this means that the coldest crevices are today found closer to the summit, while their ancestral bolt-holes below grow increasingly unsuited to aestivation. So the survival of the moths rests not just on the mountains, but on the atmosphere in which the mountains stand.

The vanishing of the bogong moth betrays an invisible and global threat, yet it is because the moths were once so abundant that their loss is palpable to us. That multiple populations of moths coalesced in the mountains, rather than remaining dispersed over a vast landmass, served to dramatise their absence. Empty caves are an image the mind can readily latch onto. The fact that the bogongs once proliferated in places of work and residence, too close for comfort, also made them memorable. Would their disappearance have been overlooked were they a less numerous, more cryptic species?

What do we stand to lose when we lose the moth? The answer is many-layered. If their worth is as a traction, joining worlds, then migratory bogongs need to be preserved *in abundance*. From an evolutionary standpoint, moving as a multitude permits an insect species to glut those predators it cannot evade and to absorb losses owing to adversities like bad weather. With no intrinsic defence (beyond camouflage), each bogong moth's safekeeping owes to the company of many more moths than appetite or attrition can account for. Beyond a minimum threshold, the moth gains new vulnerabilities. Some kindred insect species appear to require a set population size and density to be reached – a quorum – before they feel pushed to start their migration. What liberates these insects from stasis is not intention, or a move to action, but surrender to the vigour of massing together. Dwindling can therefore be a force like inertia. Is it absurd to imagine that, for insects and other animals, being deprived of migration has an emotional dimension? Between captivity and freedom is there an apprehension of restraint, of capture in a sphere smaller yet than desire fills? What is the name for this grief? Could it inhabit something so small as a moth?

Abundance matters, too, because rarity renders the migration's knock-on effects negligible or defunct: the migration's ecological consequences crumble where few moths undertake it. Though we customarily speak of needing to save a habitat to save a species, per those animals that move in vast numbers the opposite is also true. The preservation of habitat, its energetic balance, pivots on the transient animals that pass through it. A depletion of migratory animals can be a force as atrophying as plunging the land into darkness.

The bogong moth itself may not be beautiful, but it is a cornerstone of the beauty found in the mountains where the moths pollinate flowers and nourish alpine biomes. And yet, there is more at stake here than beauty. Upland ecosystems have downstream effects. The Alps are sometimes referred to as Australia's 'liquid lungs' for their watershed function and because they filter rain running into aquifers that supply the cities. If the bogong moth, a keystone species, does not recover, the tattering of mountain nutrient cycles may reduce water quality elsewhere.

To argue for shielding animal migration is to adopt an expansive definition of vulnerability, for it means not just protecting animals – safeguarding their mere existence – but maintaining their ways of being in the world, and in relationship to one another. It can also mean attending to the commonplace over the seldom seen; to the maintenance of the unremarkable. Yet only the most cold-eyed ecologist would view an animal's worth as exclusively a matter of tabulating benefit to its surroundings, dependents and consumers. The moths don't just connect ecosystem to ecosystem, they connect people to people, and people to the past. Indigenous owners of the moth aestivation grounds are cognisant of the moth as a talisman for the Country it enlivens, and of its pathway through the world as a more durable kind of knowledge than anything illuminated by data. One thing we all stand to lose if the bogong moth disappears for good is a feeling for our own smallness and impermanence – for even as each moth is turned about in less than a year, the flow it belongs to is ancient.

On a morning not long ago, as the city began to stir out of its own long inertia, I drove to Melbourne Zoo to visit the Butterfly House. In the queue corralled along the zoo's perimeter people fumbled to pull up vaccine passports on the government app, and zoo workers implored ticket holders to observe social distancing by keeping 'the length of a kangaroo' between groups: an interval soon collapsed by impatience. As the clouds flew off overhead, I wondered if the animals penned inside heard the hum of the crowd, and whether that noise aroused anxiety or anticipation in them. Had any of the zoo's creatures worried over where their spectators had gone and why they had disappeared for so long?

The Butterfly House is steamy, with rings of melon on wire suspended from the trees and hexagonal feeding tables set with plastic florets, where sugar water is set out. Around 450 Lepidoptera quiver on vegetation, or on the feeding tables, and thresh in the air. Almost all are butterflies, though there is one day-flying moth species among them – the Hercules moth, a few saucer-sized individuals very far from the tropics where they usually make their home.

There are no bogongs. To date, I have not seen one. So far this year, the mountains have not either. Yet it is hard to put into words the upswing of emotion I experience in the Butterfly House nonetheless. I can't help it. To be surrounded by so much life this fragile and ornate is overwhelming. An older man is frozen mid-step by a butterfly alighting on his forehead, on the spot notionally known as the third eye. A bygone name for Lepidoptera is *psykhe* (psychê), a term that later denoted the soul. Something moth-sized has taken flight within my body, a ricocheting brightness with an autonomy all its own, a lightness that had eluded me all through the long second half of the year. As though these vivid insects, their presence, have broken something heavy within me into parts and made it available to catch on a breeze.

Bogong moths are not social insects. Social insects, such as honeybees and termites, have a hierarchy and distribute roles to different individuals. What the moths are is massively *gregarious* – and what they are losing now is their coming together. Denied to the moths is the momentum to densify, to persevere high in the overworld, and endure ever more infernal summers by resting, wingtip to wingtip. And yet, some of the planet's most isolated insects have been restored by human effort. The Butterfly House ordinarily displays one non-Lepidoptera insect at intervals, in a terrarium brought out from enclosures not open to the public: the Lord Howe Island stick insect, a bug also called the 'tree lobster' after its size. Considered extinct since 1920, a breeding colony of just 24 Lord Howe Island stick insects was discovered clinging beneath a single shrub on a crag of rock, a sea mountain called Ball's Pyramid, in the Pacific Ocean, over 600 kilometres north-east of Sydney. The two stick insects (the zoo named them Adam and Eve), brought to Melbourne by their discoverers, have since given rise to over 14 000 offspring. I'm told they are not on display today because people too easily forget the social distancing protocols when they gather around to view them.

Are we in time to double back, to save the bogongs? The revival of the moth won't be so easy to secure. On Lord Howe Island, the stick insects were killed off by invasive rats: a localised catastrophe. The plight of the bogongs, on the other hand, is entangled with the global problem of climate change. All migratory species depend on a series of habitats, sequenced conditions in those habitats, and the transitional spaces that provide passage between them. Such species are contingent on a far greater domain, replete with many more resources, than is needed to supply the individual. A single insect may live off a tablespoon of nutriment. A dozen may occupy a shoebox without antipathy. And though a bogong could survive in the Butterfly House, the fate of its species hinges on weather patterns that encompass the globe. And yet, hope resides in the insect. Having evolved to produce hundreds of offspring (over a strategy of cosseting a few), each generation has the potential for mass depletion, but also replenishment. One adult can lay 2000 eggs. In the wake of the IUCN listing, conservation advocates are now pushing for improved moth habitat within agricultural lands, offset by subsidies similar to those offered to some EU farmers for engaging in bee-friendly growing methods.

My gaze returns to the butterflies when one lands on the back of my hand and pauses there, closing and unclosing like a pamphlet of inscrutable information. A voice piped over the speakers warns visitors not to try to stroke the insects or grab for them. *Even with the greatest of care*, it says, *our touch can hurt them.*

SPACE COWBOYS

Alice Gorman

Humans have been sending rockets into outer space for 65 years. For much of this time, we've thought of space as a vast empty nothingness, a cold void that requires all our ingenuity to survive. We never thought that outer space might need protecting from us.

In 1957 there were just three satellites orbiting Earth. Now there are more than 4800, and a frightening amount of space debris. One estimate is more than 37 000 objects over 10 centimetres in size, and tens of millions of smaller particles. In weight, space junk is the equivalent of 10 million cane toads. A collision with debris – travelling at an average speed of 7 kilometres per second – can cause spacecraft to break up or fragment, creating even more debris in the process.

Options for cleaning up this mess are limited. At the moment, the only method is waiting for objects to fall back into the atmosphere and burn up. Many of the technologies being developed to actively remove debris involve tipping it back towards Earth too. Effectively, the upper atmosphere has become a planetary-scale garbage incinerator.

However, burning spacecraft materials creates particulates of soot and alumina, which can cause ozone molecules to disintegrate. The ozone layer protects life on Earth from the Sun's savage UV radiation. This might have been manageable at previous spacecraft re-entry rates. With the launch of the megaconstellations, such as SpaceX's Starlink, the problem is going to get worse. By 2030, there may be as many as 60 000 satellites in Low Earth Orbit. Some will become junk that stays up there, increasing collision risks, while others will fall out of orbit and burn, increasing the amount of alumina in the upper atmosphere.

The Moon is at risk too. Space agencies are partnering with private companies to develop the technology to return to the Moon, and stay this time. The isotope helium-3, rare earth elements, water ice and various other minerals are the target of commercial exploitation to supply Earth, support a lunar economy and perhaps provide resources to go on to Mars. Some argue that moving mining into space will reduce the environmental impacts of extractivism on Earth. But doesn't this just mean transferring the same impacts to the Moon?

The Moon is often thought of as a dead world. This is a matter of perspective. The Moon has a very fragile and tenuous 'exosphere' of gases. It's seismically active with moonquakes occurring regularly. It also has water cycles, which we are only just beginning to understand.

Thanks to images from numerous lunar missions, including close-ups of the lunar surface taken by the Apollo astronauts, the Moon is the celestial body most familiar to us. Lunar landscapes have their own distinct aesthetic and environmental values. Because the lunar surface is not constantly renewed by plate tectonics, the history of the solar system is inscribed in its dust, craters and lava fields.

Rocket transport and infrastructure such as mining installations, power plants, waste management and habitations will inevitably impact these landscapes. The lunar surface could be 'terraformed' into an industrial landscape. Rocket exhaust could loft the fine lunar dust into orbit, creating a cloud around the Moon. The albedo, or the way light reflects off the surface, may be altered. We wouldn't likely see this with the naked eye, but satellites and telescopes will be able to capture these changes. And soon there'd be junk in lunar orbit too. Without an atmosphere to burn it, it will crash directly onto the lunar surface.

Current mining plans are focused on the water ice trapped in the Permanently Shadowed Regions (PSRs) of the lunar South Pole. Here, 2-billion-year-old shadows in deep craters shelter lakes of ice.

PSRs are rare in the solar system; to our knowledge they occur only on the Moon, Mercury and the dwarf planet Ceres in the asteroid belt. The ice will be used to make rocket fuel and provide water and oxygen for industry and to sustain human life. And, of course, there are the environmental impacts that we don't yet know about.

This highlights a broader problem. There is no environmental management framework for space. Without living ecologies, many don't see space as an environment to begin with, and so believe there are no environmental values worth protecting: space is just a set of resources for humans to use. Ironically, the 1960s images of Earth taken by the Apollo missions were pivotal in creating a new understanding of Earth as a fragile environment that needed our care. The environmental movement in space is 50 years behind the thinking on Earth.

What are the solutions? Right now, it seems there are two positions. One is to stay on Earth, sort out our problems, and give up space travel in order to avoid wreaking the same environmental havoc we have on Earth on other orbits and planets. The other is the current trajectory: a form of space capitalism in which science plays a secondary role in supporting commercial exploitation.

But these are not the only options. The Australian eco-feminist philosopher Val Plumwood outlined an approach to the environment she called 'co-participation'. In this approach, the space environment would be treated as an active agent whose integrity must be considered, rather than just a resource to be used. The ideal endpoint is a 'mutual flourishing' instead of one entity being sacrificed to the needs of the other.

One way of achieving this for the Moon might be to grant it legal personhood, as proposed by space scholars Eytan Tepper and

Christopher Whitehead. In this model, trustees would represent the interests of the Moon to ensure that human activities do not compromise its environmental integrity. This is a paradoxical position, of course, but holds promise as a way of breaking out of the binds of space capitalism.

Although the Outer Space Treaty of 1967 says that space is the common province of all humanity, in reality some nations derive more benefits from it than others. These are the same nations who cause the most damage.

In the space world, there's a lot of discussion about creating equal opportunities for 'emerging space nations', but this doesn't fix the problem of inequalities created by a long history of colonialism. The 'emerging space nations' are being forced to play by the same rules that excluded them from space to begin with, and which lead to the potential 'tragedy of the commons' in Earth orbit.

The major spacefaring nations and wealthy space corporations should be shouldering the responsibility for causing environmental harm in Earth orbit and the upper atmosphere. With the Moon, there's still a chance to get things right. But it will take new ways of thinking.

✽ *Galaxy in the desert*, p. **49**
Dark skies, p. **245**

WHERE GIANTS LIVE

Belinda Smith and
Alan Weedon

A couple of hours' drive from Melbourne, you'll find giants.

On a winter's day, they fade in and out of the gentle swirling rain, seemingly melting into the mist before abruptly bursting into view once more.

These ghostly mountain ash forests, in an area known as Victoria's Central Highlands, were home to Gunaikurnai, Taungurung and Wurundjeri peoples for tens of thousands of years. Some towering individuals extend their statuesque trunks almost 100 metres towards the sky.

Mountain ash can live up to 500 years. They are Earth's tallest flowering plants, and one of the tallest tree species on the planet. It's no wonder the second part of their botanical name, *Eucalyptus regnans*, means 'ruling' or 'reigning'.

The tallest living mountain ash we know of is in southern Tasmania – a behemoth nicknamed Centurion that stands a bit over 100 metres from forest floor to foliage crown.

In the forest near Marysville in Victoria is a mountain ash known as the elephant tree. It's not as tall as Centurion, but is still impressive, with a trunk that measures 13.6 metres around its base. And it's old – about 400 years. That means when European colonisers first sailed from Bass Strait into Port Phillip at the beginning of the 19th century, the elephant tree was already nudging 200 laps around the Sun.

Such enormous evergreens appear rock solid; capable of

withstanding whatever the years throw at them. But their future, even in the near term, looks shaky.

These forests are one Australian ecosystem that might collapse in the next decade under a warming world, according to an IPCC report released in March 2022. And there's more to lose than a few old trees.

Forest-dwellers at risk

David Lindenmayer knows the area well.

Today, he's a forest ecologist and conservation biologist at the Australian National University. But back in 1983, he was working in the mountain ash forests near the town of Marysville as a fresh-faced science graduate. His job was to take a census of sorts, but instead of people, he surveyed Victoria's state animal, Leadbeater's possum (*Gymnobelideus leadbeateri*).

He wanted to know how many there were, where they lived and what aspects of their habitat were crucial to their survival. This was no small task, as the pocket-sized marsupials are most active at night and prefer to hang out high in the treetops.

Leadbeater's possums had already come back from the dead once, in a way. In the 1930s, they were thought extinct – driven to their demise by farming and fire. But in 1961, they were rediscovered in the Central Highlands' mountain ash.

Professor Lindenmayer set to counting in 1983, and eventually broadened his census to other resident mammals.

'There's a beautiful assemblage of possums and gliders in that forest, including greater gliders, yellowbelly gliders, mountain brushtail possums,' he says. 'When I first started working, about one in every three large old trees had a possum or a glider in it. Now the numbers are about one in every eight or every nine trees.'

When he'd go out spotlighting in the early days of his career, the most common species he'd catch in the light was the greater glider. That species has declined by around 80 per cent over the past 20 years. In July 2022, the greater glider was reclassified from 'vulnerable' to 'endangered' at a national level.

And the little Leadbeater's possum?

'They've declined by about 50 per cent,' Professor Lindenmayer says.

Crucial to the survival of these and other forest-dwellers are stands of old-growth forest that contain trees more than 120 years old. Older trees are full of hidey-holes where all manner of animals – not just possums but a whole bunch of birds too – make their nests and feed on leaves and buds.

Fewer big old trees mean fewer places to live and less food. And big old trees are already few and far between. Back in the 1980s, about 25 000 to 30 000 hectares of old-growth mountain ash forest stood in the area, Professor Lindenmayer says.

'Now it's 1886 hectares. The abundance of big trees has really dropped through the floor.'

Guardians of the water supply

Animals further afield – Melburnians – rely on old-growth forest too.

That's because most of Melbourne's water supply comes from the mountains to the east of the city where the mountain ash live. As snow and rain fall on the ranges, it trickles through the soil to the water table underground, and burbles out again on the surface as springs. These springs feed creeks, which feed rivers, which feed reservoirs, which feed into our homes.

So what do old-growth mountain ash trees do for our water supply? It's more about what they don't do.

While mountain ash keep growing throughout their whole life, their twilight years are spent growing far slower than when they were saplings. They shoot up around a metre every year during their first 70 years or so. This needs a lot of water, so some of that reservoir-bound rainfall is slurped up by thirsty roots instead. Keep as much old-growth forest intact as possible, and Melbourne's water supply is protected too.

Why are these old trees vulnerable?

You don't have to look far to see where threats to these forests come from.

The narrow shoulder of the winding road from Marysville to the elephant tree is peppered with signs warning of logging trucks. Black scars of fires from years gone by peek from beneath new growth. Either frequent burning or logging on their own is bad enough, Professor Lindenmayer says – but together, they spell disaster.

Let's start with fire. Mountain ash and its slightly smaller cousin alpine ash (*Eucalyptus delegatensis*), also found in this area, are 'seeders' and need fire to reproduce. After a bushfire, the trees sprinkle thousands of tiny seeds on the ground. Mountain ash need to be around 20 years old before they make seeds – and that's normally fine. This is a cool, wet part of the country, and fires tend to be spaced out enough to let young trees mature to reproduction age.

But climate change has thrown a spanner in that balance between death and rebirth, and is driving fires more frequent than the trees have evolved to handle.

Young, recently logged forest is more vulnerable to fire for a few reasons. First up, the stuff that's left on the forest floor after the big trees are taken away – branches, leaves and the like – can be fuel for fire, Professor Lindenmayer says.

'Our data also shows that very young forest is much warmer, and more prone to extremes of temperature. Plus wind speeds are higher, which is a really important part of fire dynamics.'

And then there's the baby trees themselves, which are more flammable than their older siblings.

If fires wipe the mountain ash slate clean more than two or

three times a decade, and the forest doesn't have time to recover, they disappear, and other trees take over.

It's not just young trees that are vulnerable to burning. The biggest and oldest mountain ash trees can survive fires, but they can only take so much.

Repeated fires – and not even particularly intense ones – can kill off the big old trees, says Lachie McBurney, a research officer with ANU who lives nearby and monitors the ash forests. If a fire whooshes through a patch of forest, it might not kill big trees, but can leave a blackened scar on their base.

'When you get a fire scar in the bottom of a big tree, that's like a chimney door for when the next time a fire comes through,' Mr McBurney says. 'So even if you get a really light burn [next time], like a fuel reduction burn ... that'll actually turn that into a chimney and kill that tree. Even a massive giant we think, because it's big, it should stand for longer, but that's not actually the case.'

Throw in a commercial logging regime that harvests trees under 50 years old, and the forest as a whole remains younger, Professor Lindenmayer says.

The forest's age also affects the amount of carbon it locks up. Mountain ash forest can hold 1900 tonnes of carbon per hectare – up to 10 times more than tropical rainforest – but it needs to be old growth. So along with using more water, young forests also store less carbon than their older counterparts.

'And the more of these catastrophic fires we have, the more we risk going into a downward spiral,' Professor Lindenmayer says.

Respect for arboreal elders

As the world warms, the outlook isn't great for mountain ash.

To give the forests the best chance of survival, we need to stop logging them immediately, Professor Lindenmayer says.

The timber industry has a long history in the Central Highlands. By the 1920s, these ash forests supported around

120 sawmills. In the industry's heyday, thriving communities, complete with schools, were scattered throughout the forest.

These days, all but a couple of houses are gone, torn down or destroyed by fire. Abandoned machinery sits rusting among the trees. What also remains is an enduring effect on Melbourne's water supply.

Professor Lindenmayer and his colleagues calculate the city loses 15 billion litres of water each year thanks to logging in the city's most important drinking water catchment, the Thomson Catchment. That's enough water for a quarter of a million people.

And if logging in the catchment continues at its current rate, the equivalent of 600 000 people's water will be lost from Melbourne's water supply every year by 2050.

'As time has gone on, it's become clearer and clearer that the level of pressure on the mountain ash system is just unbearable,' Professor Lindenmayer says. 'It's not possible to keep logging the forest the way it's being logged, and to have so much fire in the system, and to expect any of the forest species to be able to survive – including the mountain ash forest itself.'

It's easy to feel insignificant when gazing up at these centuries-old giants.

But their fate, for better or worse, lies with us.

✻ *This magnificent wetland was barren and bone-dry. Three years of rain brought it back to life*, p. **28**
Point of view, p. **121**

TALARA'TINGI

Felicity Plunkett

To lie among flames, to come through fire
in the shape of a star – (o hope) –

feathery. To push pink past char, woolly
across heath and open forest, clumping

to blush, jump bruised blue eucalypt
–hazed mountains lying grazed – (o hope) –

To be given a man's name – *forsythii* – not to be
named, to be nameless, to know

what it is to be spoken over, spoken
of, pressed. To open underground – (o hope) –

To come back – hailed by lyre, by whip – from
catastrophe – after the flicked

cigarette, flash of hand and stripe of dry
lightning. To follow the drowning – (o hope) –

carrying colour like a blanket. To labour in fire
'furnished with rays' – *actinotus* – homeless. To go

across razed borders under a burgeoning
enemy's thunder, after everything

To offer an artillery of fluffy seeds to breeze, open
velvety bracts high above cousins coasting

silver. To hold dew-nectar morning bright, to soothe
the wounded. To be wounded, to lie under disaster

– (o hope) – To make something anyway. To turn
from your moribund cradle into roseate light, into air –

– (o hope) –

❋   *This magnificent wetland was barren and bone-dry. Three years*
    of rain brought it back to life, p. **28**
    Point of view, p. **121**

THE PSYCHEDELIC REMEDY FOR CHRONIC PAIN

Clare Watson

Every day, James Close meets people living with unrelenting, insurmountable chronic pain. A pain physical therapist and research fellow at Imperial College London, Close seeks to help people who have exhausted every treatment that specialists have to offer, and he sees their desperation to regain a semblance of a normal life – physically, emotionally and socially.

Across the Atlantic, Joel Castellanos, a pain medicine and rehabilitation physician at the University of California, San Diego (UCSD), asks his patients about their pain and how it interferes with their lives. He hears how their persistent pain wrenches away their ability to work, socialise and exercise.

Looking for new therapeutic options, the two pain therapists have found themselves at the forefront of research exploring whether mind-altering psychedelic drugs might provide the relief that conventional analgesics (painkillers) cannot. Psychedelic medicines coupled with psychotherapy might 'change the way we experience pain', Castellanos says.

There is not much clinical evidence yet – just a smattering of findings across a handful of chronic pain conditions hinting at the possibility that psychedelics can alleviate persistent pain for weeks or months at a time. But a slew of pilot trials pitting psychedelics against chronic pain are starting to help the idea gain traction. These trials could provide the evidence to either establish psychedelics as a pain-relieving treatment or rule them out.

The trials build on numerous surveys and case studies that report profound reductions in the severity of cluster headaches, fibromyalgia and phantom limb pain. Researchers are also leaning on tentative evidence suggesting that psychedelics can ease the depression that often accompanies chronic pain, and on decades-old reports that psychedelics can outperform opioid medications in managing cancer pain. 'There is increasing recognition that psychedelics could be helpful in the treatment of chronic pain,' says Peter Hendricks, a clinical psychologist at the University of Alabama in Birmingham.

Despite the undeniable need for better pain relief, many pain researchers – including Close, Castellanos and Hendricks – are urging caution. Clear evidence that psychedelic therapies can quell pain is still lacking – indeed, psychedelics can sometimes worsen some people's pain. Moreover, people with chronic pain are particularly vulnerable to hope and hype. 'The potential is really grand,' Castellanos says. 'But we have to do the work.'

Mechanisms of mitigation

There are several plausible biological mechanisms for how so-called classic psychedelic drugs, such as psilocybin and LSD, might relieve chronic pain. These powerful drugs bind the serotonin 5-HT2A receptor, which is integral to the central pain responses that go awry in chronic pain.

Studies in mice show that by activating 5-HT2A receptors, classic psychedelics dampen inflammation, which is a key driver of chronic pain, and upregulate genes involved in synaptic plasticity – the brain's ability to strengthen, loosen and reorganise its connections that could, in this case, rejig pain networks. In human brain imaging studies, psychedelics seem to reset brain circuits that are perturbed in chronic pain or appear overactive in headache disorders.

'We're getting closer to seeing the rationale as to why

psychedelics may be relevant in chronic pain,' says Bernadette Fitzgibbon, a neuroscientist who studies pain at Monarch Research Institute in Melbourne, Australia. But researchers are a long way from joining the dots between the pharmacology of classic psychedelics and their apparently rapid and transformative effects on the brain networks involved in pain. And they are not close to showing whether the drugs are effective analgesics.

Nevertheless, a few small clinical trials are building on anecdotal evidence that psychedelics provide a reprieve from searing cluster headaches, for which conventional medicines provide little relief. The effects are said to last months, or even years. Intriguingly, participants in these studies report pain relief even at low doses and without the mind-altering experience that underpins the way psychedelic-assisted therapies might attenuate depression and post-traumatic stress disorder (PTSD).

The results from a double-blind, placebo-controlled trial of 10 people who experience migraines – the first study of its kind – offer the first inkling that the anecdotal reports might stand up to clinical scrutiny. The study, led by neurologist Emmanuelle Schindler of Yale University in New Haven, Connecticut, found that a single, low dose of psilocybin – small enough not to induce a trip – halved the number of days people experienced migraines, over a two-week period. 'There's nothing in headache medicine that can really do that,' says Schindler.

But long-term relief after a single psychedelic treatment is uncommon, as most people tend to report greater relief with multiple doses, Schindler says. She is running a follow-up migraine trial comparing two doses of psilocybin with one, as well as a longer, larger trial testing psilocybin against cluster headaches. In that randomised trial, which is similar to a study underway at the University Hospital of Basel in Switzerland, participants were given three doses of psilocybin (or a placebo) spaced five days apart; they then monitored their headaches and quality of life over the following two months. The results are forthcoming.

Cluster headaches, however, are a very different beast to migraine, even though both are paroxysmal disorders in which pain comes and goes. So despite the early evidence, Schindler, like other trial investigators, stresses that psychedelics might not work in other types of chronic pain, or indeed for everyone.

Complex conditions

That reality has not deterred the researchers. Pain comes in many forms, and even though chronic pain conditions have unique features, there are commonalities in which the pain-relieving potential of psychedelics lies – although that remains to be tested. The unrelenting nature of chronic pain also presents a challenge quite unlike that of episodic headaches.

Nevertheless, a pilot trial underway at UCSD is testing psilocybin's ability to treat phantom limb pain – a quintessential neuropathic pain condition that Castellanos says serves as a good test case for the use of psychedelics.

Neuropathic pain is triggered when pain-sensing nerves are damaged or severed. For amputees with phantom limb pain, excruciating pain seems to emanate from the limb that is no longer there. But psychedelics have, in some rare cases, completely eliminated phantom limb pain when combined with therapies that use visual illusions to trick an amputee's brain out of its pain.

In the randomised, triple-blind trial at UCSD, 30 amputees with phantom limb pain will receive either two high doses of psilocybin or a placebo dose of niacin, chosen because it delivers some of the buzz people might get from psilocybin, though without evoking hallucinations. Psychotherapy sessions will help participants prepare for and process their psychedelic experience, while functional brain imaging will look to see whether pain circuits are reorganised after treatment.

It is thought that psychedelics might disrupt connections between resting state brain networks that become ingrained as people ruminate on their pain or fixate on the ever-present threat of it coming back. Castellanos says that for people with persistent pain, their bodies are 'practising experiencing pain every day'. Psychedelics might be able to reset those neural pathways.

But researchers doubt that psychedelics alone will provide relief. Chronic pain is influenced by a wide range of cognitive, emotional and social factors, layered on top of sensory and bodily cues. By removing people from their pain momentarily, psychedelics might trigger a transformative experience, providing an emotional shift that creates fertile ground for further therapy. But they will need to be paired with psychological and physical therapy because psychedelic drugs merely open a window, as Fitzgibbon puts it, for psychotherapy to provide sustained benefit.

Close suggests that in the long run, combined therapies could ease pain that once seemed permanent by helping people reframe the way they relate to their chronic condition or perceive persistent pain.

Focusing too much of clinical research on using psychedelics as analgesics, Close adds, also risks overlooking the psychological aspects of pain, such as depression and anxiety, which intensify the pain experience. 'Looking at it just through a physiological lens does not help,' he says, especially when chronic pain is rooted in and exacerbated by social, cultural and political inequities. Those inequities feed into the complex web of cognitive, emotional and behavioural responses to chronic pain that overlap and interact to influence someone's perception of it. A holistic approach to pain management is therefore needed, Close says.

This is especially true for fibromyalgia, a centralised pain disorder marked by widespread body pain and fatigue that worsens when people avoid movement. Fibromyalgia typifies central sensitisation, whereby nerves in the brain and spinal cord misfire and cause pain in joints, muscles and tendons. Depression is more common with fibromyalgia than with other pain disorders, making it another interesting test case for psychedelic-assisted therapies.

Hendricks is embarking on a feasibility study to test whether psilocybin is an effective treatment for fibromyalgia. It's not the only one: an open-label clinical trial led by Kevin Boehnke, a chronic-pain researcher at the University of Michigan in Ann Arbor, is underway, and Close is leading a third study at Imperial College London.

Hendricks's trial is starting small, aiming to recruit 30 fibromyalgia patients and asking them to report changes in pain severity and other measures of quality of life over three months. Dextromethorphan, a drug that can have hallucinogenic effects similar to psilocybin, will be used as a placebo.

Meanwhile, Close's study at Imperial will capture first-person accounts of people's pain experience after taking psilocybin, and look for changes in brain activity and connectivity. Clearly, Hendricks says, there is a lot of enthusiasm around what psychedelic-assisted therapy could do for people with fibromyalgia. 'But ultimately, we have to let the data do the talking.'

Tentative steps

Even if psychedelics cannot stop pain signals per se, researchers speculate – based on limited evidence – that these drugs might still hold some potential to alleviate the depression and substance use disorders that often accompany and exacerbate chronic pain. If so, that would 'undoubtedly improve quality of life among people with chronic pain conditions', says Hendricks. 'There is room for improvements even without pain relief,' Close adds.

Multiple surveys and one longitudinal study of people who use illicit drugs have linked lifetime psychedelic use with lower incidence of opioid use disorder. Some data also suggest that psilocybin and LSD could help treat addiction to nicotine and alcohol. What's more, there is considerable physiological and neurological overlap between depression and chronic pain. Repetitive, ruminative thoughts and catastrophic thinking, in which people focus on bad outcomes, are key facets of both depression and chronic pain. Inflammation is another link between the two that extends to migraine, fibromyalgia, cancer-related pain and autoimmune diseases.

'These things are hardly separable,' says Frederick Barrett, a cognitive neuroscientist at Johns Hopkins University in Baltimore, Maryland. His brain imaging studies have detected changes lasting up to one month after a single, high dose of psilocybin in two regions of the brain, the dorsolateral prefrontal cortex and the anterior cingulate cortex, which are linked to mood management, pain modulation and pain perception.

Barrett is being careful not to overstate the potential of psychedelics to treat chronic pain; the evidence so far is nowhere near what's needed, he says. But the clinical trials that are underway, he adds, are exactly the kind of tough test that researchers should be putting psychedelic therapies through as they carefully evaluate the potential benefits and risks. The next steps include figuring out in larger trials how often psychedelic medicines need to be delivered, at what dose, and what supportive care is needed for each chronic pain condition.

Understanding the obscure mechanisms by which psychedelics evoke such powerful and durable effects is another key research goal, and could yield new insights about existing pain-relief medications. 'There's potential to learn a lot about mechanisms of chronic pain, and what needs to happen centrally in the brain and spinal cord when patients get better – or when they don't,' says Castellanos. Even without fully elucidating the underlying mechanisms, the profound and lasting effects of psychedelics reported by some patients suggest that these drugs are not merely covering up the symptoms of pain disorders but targeting the root causes of chronic pain, Schindler says. And if headache disorders are anything to go by, she adds, relieving chronic pain might not always require a full-blown psychedelic experience, as low doses seem to be effective.

Researchers might be keen to validate anecdotal reports from people with pain, Close says, but they still need to design rigorous studies and be honest about the outcomes. Otherwise, he warns, if pilot trials are rushed, the risk is 'you end up with a treatment that might help some people but we don't know who they are, so you have lots of people trying it who it doesn't help at all.'

Careful not to repeat the mistakes that led to the opioid epidemic, researchers are edging forwards, one small trial at a time. Perhaps in the near future, psychedelics could help some people cast off or delve into their pain, re-engaging with their physical bodies that they have long dissociated from, and begin at last to heal.

✱ *A whole body mystery*, p. **113**
  Isolation, p. **226**
  Do we understand the brain yet?, p. **262**

A MYSTERY OF MYSTERIES

Fiona McMillan-Webster

One day a few years ago, Alfred was sitting by himself, eating some fruit and generally minding his own business. If Alfred knew he was being watched, he didn't seem to mind – it wasn't the first time he'd been shadowed like this. As he enjoyed his snack, those following him crouched in their hiding spot a short distance away, taking notes, recording footage, observing how he held the fruit and precisely how he ate it. They stayed with him all day, and the next, and the one after that: about five days in all. If he rested, they rested. If he moved, they moved, too. And during these curious five days, whenever Alfred took a moment to relieve his bowels, this prompted a quiet flurry of activity by those surveilling him. The location was noted with great interest, as well as the date and the hour. As Alfred moved away, out came the collection bag. Alfred, you see, is a Bornean orangutan, and his observers are part of the Tuanan Orangutan Research Program based in Gunung Palung National Park on the Indonesian island of Borneo. Their goal is to study the behaviour and ecology of critically endangered orangutans like Alfred and determine, among other things, whether these orangutans are effectively gardeners.

The answer, as it turns out, involves the rather interesting mathematics of orangutan poo.

Boston University researcher Andrea Blackburn wanted to know just how good Bornean orangutans are at dispersing seeds. Over the course of more than a year, she and her colleagues followed dozens of orangutans, including Alfred, for up to five days each. It was an exercise in patience and the efficacy of bug-repellant, and it resulted in the collection of 733 faecal samples. According to Blackburn, around 70 per cent of those samples contained intact seeds. Sometimes they found thousands of tiny seeds in a sample, other times it was one or two big seeds, says Blackburn. 'There was this pretty big range of diversity in terms of the seeds that we found.' In short, over a five-day period, each orangutan dispersed an average of around 30 seeds representing up to nine genera of plants. On average, the seeds were dispersed 450 metres from their parent trees, with some ending up more than 2 kilometres away.

Importantly, not only were the seeds still able to germinate, but in most cases they were also better at germinating than the same species of seeds that had not taken a ride through an orangutan's digestive tract. The plants that seemed to benefit the most from this gut treatment were the fruiting canopy trees *Alangium* and *Tetramerista glabra*.

In a sense the orangutans do act like gardeners, says Blackburn. 'In terms of a conservation perspective, each individual is pretty important,' she tells me. Orangutans are the largest arboreal frugivores and they swallow seeds of many different sizes. In addition, says Blackburn, 'Orangutans can open these really big fruits that often have hard husks, like durians and other fruits, that other smaller frugivores and smaller primates cannot get into. So I think they're dispersing those seeds – even if they're not necessarily swallowing them, they might be just dropping the seeds and other animals can come and move them or interact with them.'

With the help of orangutans, these canopy trees are able to expand their 'seed shadow', which is the range in which a plant's seeds are dispersed and will germinate. At the same time, while the seeds are temporarily inside an orangutan, they cannot be predated upon, and therefore destroyed, by other animals.

Orangutans are not unique in their role as unwitting gardeners, which is an important part of the evolutionary strategy of seed plants. Paleobotanist Andrew Rozefelds points out that seed dispersal by animals can be critical to plant survival. Seeds that stay close to a long-lived parent tree will find themselves immediately in competition for resources, he says, and if they can't move, they're stuck. So key animal species play a vital role by moving seeds around within a habitat. 'By moving a seed away from a host tree,' says Rozefelds, 'it's got a good chance of surviving somewhere else.'

Studies have shown that up to 90 per cent of seeds in tropical regions are dispersed by animals, largely via ingestion. This process, called endozoochory, plays a major role in seed dispersal globally. If you have ever eaten a wild blackberry or blueberry while wandering through a woodland, there's a good chance it was grown from a seed that had, at some point, passed through the fundament of a wild animal.

Berries are curiously popular across a wide swathe of the animal kingdom, from birds and rodents to deer, bears, even wolves. In a 2018 study of eight wolf packs in Minnesota, animal ecologist Joseph Bump and his colleagues found that blueberries comprised up to 83 per cent of the wolves' mid-summer diet. Meanwhile, other studies show that in a season where sarsaparilla berries, cherries, raspberries, juneberries, blueberries and more are in abundance, a single black bear, or *Ursus americanus*, can eat as many as 30 000 berries in a day. Brown bears are similarly prolific when it comes to consuming fruits and dispersing seeds, with studies showing that *Ursus arctos* scats (poo) often contain tens of thousands of viable seeds.

In 2018, ecologists Laurie Harrer and Taal Levi from Oregon State University studied the eating habits of brown and black bears in south-eastern Alaska and discovered these bears had a staggering fondness for the berries of the devil's club plant, a woody shrub native to the coniferous, old-growth forests of North America's Pacific Northwest. The bears were observed consuming around 30 berries per second, which equates to just over 100 000 berries per hour of foraging. Based on this, the researchers estimated the bears were dispersing around 200 000 seeds per square kilometre per hour. These animals are well-known seed dispersers, and, much like Blackburn's findings with orangutans, gut treatment seems to have a germination-enhancing effect on many of the seeds they ingest.

Vertebrates, especially mammals and birds, are major seed dispersers in a wide range of habitats. The list is long, varied and sometimes a bit weird, and what follows is by no means exhaustive. To begin with, there are shorebirds that disperse seeds of flowering plants and fruit trees to distant islands; migratory swans that spread pondweed seeds across continents; and mallards in Hungary that carry the seeds of elderberries, figs and bittersweet nightshade to the edge of the Black Sea. There are bluebirds that disperse myrtle seeds across South Carolina, and African and Asian hornbills that disperse a wide variety of plant species. Bats, which account for roughly 20 per cent of all mammals, are prolific seed dispersers. African elephants, which quite enjoy a nice fig or the fruit of the baobab tree, can disperse seeds up to 65 kilometres from where they feasted.

As suggested by the actions of bears and wolves, some carnivores are surprisingly proficient at seed dispersal. Coyotes, it should be noted, seem to have a taste for hackberries and persimmons. There are also many seed dispersers in the Mustelidae family, which includes weasels, racoons, otters, martens and wolverines – carnivores all.

One of the oldest living family of terrestrial carnivores, the Viverridae, the most well known of which are civets, are pretty good at it, too. Asian palm civets are native to Indonesia, and have a cat-like appearance and something of a coffee habit. They enjoy eating ripe coffee cherries, and the undigested coffee beans end up in their droppings. This should be a win-win: the coffee plant widens its seed shadow and the civet gets a nice fruity snack. But their mutualism is increasingly being interrupted due to the rise in popularity of *kopi luwak*, a drink made from the coffee beans found in civet dung. Costing up to $100 a cup, it's the most expensive coffee in the world. Sadly, this luxury status has led to the capture and caging of many civets, and, as some experts argue, all for coffee that doesn't even taste that good anyway.

Herbivores, as you might expect, are excellent seed dispersers and vitally important to many ecosystems, especially the grazing ungulates: horses, giraffes, hippopotami, tapirs, all the bovines, and pretty much any other mammal with hooves. Lowland tapirs (*Tapirus terrestris*) consume more than 350 plant species and are particularly good at dispersing their seeds in the Amazon. Even reptiles get in on the seed-dispersal act, from tiny skinks to enormous, plodding tortoises. Seed scientists Si-Chong Chen and Angela Moles have noted that Galápagos tortoises not only swallow seeds but can hold them in their guts for up to 33 days. As Chen and Moles explain in an article in *Australasian Science* from March 2016, tortoises might move slowly, but they can cover a lot of ground in just over a month. As a result any seeds they've consumed can be spread quite far.

Chen is a researcher at the Royal Botanic Gardens, Kew in the United Kingdom, while Moles heads The Big Ecology Lab at the University of New South Wales, and together they compiled a global database of more than 13 000 animal–seed interactions across all vertebrates. The majority of animal species on the list were mammals and birds, plus a number of reptiles, and there were a few entries that surprised even Chen and Moles, such as armadillos, aardvarks and, most curiously, some species of fish.

Yes, fish.

The Pantanal in South America is the largest tropical wetland in the world. Spanning more than 200 000 square kilometres, this seasonal floodplain sprawls across the central western edge of Brazil and crosses into Bolivia and Paraguay. In an article she wrote for the World Wide Fund for Nature (WWF) in 2010, the late Brazilian journalist and conservationist Geralda Magela likened the Pantanal to 'a huge soup plate that slowly fills up with water and overflows in the rainy season, gradually empties during the dry season and then starts to fill up all over again'.

The Pantanal's enormous ecosystem is home to the highest concentration of wildlife on the South American continent. And the water available via a vast network of streams, rivers and lakes is central to the survival of thousands of plant species on which that wildlife depends. Some, such as the native tucum palm, release their fruits during the wet season, often in the midst of flooding. As terrestrial animals move to higher ground, fish take over many of their seed-dispersal duties. During this time, pacu (*Piaractus mesopotamicus*), a species of freshwater fish with disturbingly human-like teeth, feed on fruit and nuts that fall into the floodwaters. The seeds are often swallowed in the process and later deposited upstream. Ecologist Mauro Galetti from Brazil's Universidade Estadual Paulista, along with his colleagues, discovered that the tucum palm relies heavily on pacu for this service. Moreover, as they report in the journal *Biotropica*, the pacu acts as the primary seed disperser for many species of fruit-bearing plants found in the Pantanal. In fact, at least 43 species, representing a significant amount of the Pantanal's tree diversity, rely on fish for seed dispersal.

Then there is the strange case of a seed-dispersing amphibian. In 1986, researchers were studying a coastal ecosystem on the Brazilian coast, not far from Rio de Janeiro, which involved collecting a local species of tree frog, *Xenohyla truncata*. One morning, while transporting some of these frogs to the laboratory, one of the researchers noticed something unusual: there appeared to be fruit seeds in the frogs' droppings. Amphibians, including tree frogs, are largely insectivorous, relying on a diet of insects. Many amphibians even branch out into more general carnivore territory to eat spiders, worms and other invertebrates, and some will even dine on the occasional lizard or mouse. As a rule, though, amphibians don't go for fruit. But here were these *X. truncata* frogs pooping out fruit seeds. The team collected more frogs and found that they were indeed frugivorous – they regularly supplemented their diet with the small fruits of five species of local native plants, then shed the seeds around their immediate ecosystem.

Not all animal-dispersed seeds are 'gut treated', of course. Many simply adhere to an animal as it brushes past, enabling it to be transported long beyond the time limits imposed by a few healthy bowel movements. A seed snagging on a bird's feathers could wind up anywhere within the same forest or on another continent entirely. In a 2006 study, Spanish researchers Pablo Manzano and Juan E Malo placed four species of herb seeds on the wool of merino sheep, which then travelled along an ancient transhumance route spanning several hundred kilometres from northern Spain towards Extremadura, near the Portuguese border. A significant number of the seeds made it to the other end, with many more presumably dispersed along the way.

But lest we give vertebrates all the glory, it is important to realise that invertebrates also play a major role in this ecosystem engineering. The act of seed dispersal by ants is called myrmecochory and it contributes to the dispersal of at least 11 000 species of angiosperms – or, to put it another way, 4.5 per cent of all flowering plants. Chen tells me that when animals ingest seeds and disperse them elsewhere 'it's not for good intentions – they want the food'. She explains that even when it comes to ants, the seed must provide a reward. For this reason, seeds dispersed by ants usually possess a structure called an elaiosome. Under a microscope, it looks almost like an unnecessary flourish, an extra bit of fluff or an errant appendage, but elaiosomes are rich in fats and proteins, making them incredibly enticing to the ants. Because they are either sticky or structured such that they act like handles ants can grasp onto, elaiosomes make it easier for ants to transport the seeds to their nests. Although the seed is lipid-rich, the ants may only eat the elaiosome and then discard the seed, says Chen. But it's still viable and because the soil around ant nests tends to be rich in organic matter and nutrients, it enjoys a better chance of germination.

Ants are the most studied of invertebrate seed dispersers, but by no means are they alone. In Asia, hornets disperse the seeds of the flowering *Stemona tuberosa*, and in New Zealand there are native grasshoppers called weta that eat fruit and defecate seeds in much the same way small mammals do. Indeed, some weta are about the size of a mouse and weigh as much as 70 grams. Their unsettling hugeness is how they got their name. 'Weta' derives from the Māori word *wētāpunga*, which means 'the god of ugly things', although these days the grasshoppers are more often described as gentle giants. Earthworms can be helpful, too, by dispersing seeds not horizontally, like other seed dispersers, but vertically. The worms remove seeds from the surface and scatter them deeper into soil where they have access to nutrients and are afforded protection from seed predators.

This raises an important point: not all seed eaters are good seed dispersers. While many birds do indeed disperse seeds far and wide, birds and quite a few small mammals and insects are seed predators. This means they tend to digest the seeds they eat, for the fats, carbohydrates and proteins within, and the seed is destroyed in the process. That's not a good outcome for the plant, and so some plants have evolved seeds with harder coats, or sharp, barbed fruits, or found other ways to dissuade predators.

Now, a tough exterior might afford some protection, and even enable dormancy, but it presents an entirely different logistical problem for the plant: how does a tiny plant embryo break out of something like that? It is, so to speak, a tough nut to crack. Many plants arrived at the elegant, albeit smelly, solution of temporarily hiding their seeds in the digestive system of a different animal. For the seed, there are multiple benefits: (1) the seed predator can't find it; (2) the animal provides seed-dispersal services by moving about; and (3) the digestive enzymes and acids can breach the seed coat just enough so as to make germination a whole lot easier when the seed emerges at the other end.

Oh, and (4) instant fertiliser.

And so, seeds did not just evolve to take advantage of gusts of air, or to tumble easily down the slope of a hill, or float on a stream or an ocean current or survive a fire. Many seeds are the way they are now due to millions of years of interactions with animals. Some became incredibly tough, or were protected by sharp, pain-inducing fruits and cones and other off-putting coverings, while others were packaged up in fleshy, tasty fruits, full of sugars, fats, proteins and nutrients. Plants evolved pigment molecules and aromatic compounds so they could be seen and sniffed out by seed dispersers when the seeds were properly developed and ready for spreading. Indeed, there is mounting evidence that these cues are so specific that they can relay the nutrient content of the fruit, attracting, for example, seed-dispersing birds that prefer high lipids. This is why ripe fruit looks, smells and even tastes different to unripe fruit. It's why tiny elaiosomes are like an irresistible take-home meal complete with handles. Some seeds developed barbs and hooks to latch onto passers-by. Burrs are so effective in their hook-catch mechanism that they inspired the invention of Velcro.

Animals changed the way plants present their genes to the big wide world, but mutualism, by definition, is never one-sided. Seeds changed animals, too. It was a matter of survival. Over time, a species could either take advantage of the energy and nutrients provided by available seeds or seed-containing fruits and grasses, or die out. So, through evolutionary processes, animals honed their senses and their anatomies, competing and compromising not only with the plants but also with other animals. Jaw and teeth structures changed to crush hard seed coats, and digestive systems adapted as well. So did visual systems, olfactory neurons, the structure and arrangements of tastebuds, and even memory, as well as the neurological circuitry that links incoming stimuli with reward systems – for example, the way sugar drives a spike in dopamine in our primate brains, whereas the taste of a clump of dirt does not. Sometimes this took millions of years. Sometimes it took a surprisingly short period of time. If we look closely at the right species, we can witness evolutionary arms races unfolding in real time.

The weather was fine and warm on the morning of 17 September 1835 when a 26-year-old naturalist by the name of Charles Darwin disembarked from the HMS *Beagle* and stood on the dark, rocky shore of a small, volcanic island in the Pacific. He wasn't exactly blown away by its beauty. 'Nothing could be less inviting than the first appearance,' he later recounted. 'A broken field of black basaltic lava, thrown into the most rugged waves, and crossed by great fissures, is everywhere covered by stunted, sun-burnt brushwood, which shows little signs of life.' The *Beagle*'s captain, Robert FitzRoy, was even less enamoured of the scene, describing it as 'a shore fit for Pandemonium'.

They had landed on Chatham Island, now known as San Cristóbal, the eastern-most island of the Galápagos archipelago which straddles the equator some 960 kilometres west of Ecuador. The crew of the *Beagle* was in the process of finalising a survey of South American coastlines and nearby islands and was preparing to venture across the Pacific.

As Galápagos was equatorial, and therefore firmly within the tropics, Darwin had expected it to be a little more, well, tropical: 'I saw nowhere any member of the Palm family, which is the more singular as 360 miles northward, Cocos Island takes its name from the number of cocoa-nuts.' Indeed, there were no shady palm trees here or lush ferns. Instead, the islands presented Darwin with a series of alien landscapes, which he variously described as 'desolate', even 'sterile'. 'All the plants have a wretched, weedy appearance,' he wrote. 'I did not see one beautiful flower.' There were no frogs, either, which he found puzzling, although he did observe some giant, lumbering tortoises and a variety of lizards. Captain FitzRoy took a particular dislike to the 'hideous' iguanas, of which he said 'few animals are uglier'. Darwin found them intriguing but nonetheless referred to them as 'imps of darkness'. Darwin also noticed – offhandedly, at first – that the islands were home to a large number of 'dull-coloured birds'.

Darwin spent five weeks on those harsh, sunburned islands, observing and collecting specimens, and in so doing he had his earliest glimpses into something truly remarkable. 'The natural history of these islands is eminently curious, and well deserves attention,' he would later write. 'We seem to be brought somewhat near to that great fact – that mystery of mysteries – the first appearance of new beings on this earth.' The archipelago that had seemed at first so uninviting would irrevocably change the way Darwin viewed all living things. Those wretched weeds, those imps of darkness, those tortoises and, especially, those dull-coloured birds would provide the foundation on which Darwin built his famous theory of species evolution by natural selection.

As Darwin explored one island and then another, he observed a great variety of birds, which he identified as mockingbirds, wrens, warblers and blackbirds, as well as a number of finches. Regarding these birds – all but the mockingbirds – Darwin noticed that their beaks varied significantly, ranging from small and needle-like to large and rounded. As psychologist and science historian Frank Sulloway explained in a 1982 article in the *Journal of the History of Biology*, these observations did not trigger any epiphanies about speciation at the time because Darwin truly thought the birds belonged to entirely different genera and, in some cases, entirely different families. It was only after Darwin returned to London that the proverbial penny dropped.

Soon after his arrival back home, Darwin donated his specimen collection to the Zoological Society of London, upon which the young illustrator and ornithologist John Gould inspected them closely. What he found surprised even Darwin. There were finches, yes, but the warblers, the wrens and those blackbirds were nothing of the sort. They were finches, too. Darwin had collected

14 species of birds belonging to one genus. They were unique to the Galápagos archipelago, and, it seemed, many were unique to individual islands. Long after the *Beagle* raised anchor in the Galápagos, Darwin reflected on the diversity of those finches – and all those remarkably different beaks. He was at once puzzled and enthralled: 'Seeing this gradation and diversity of structure in one small, intimately related group of birds, one might really fancy that from an original paucity of birds in this archipelago, one species had been taken and modified for different ends.'

One might, indeed. In this statement, written years before he published *On the Origin of Species*, you can almost feel the subtle yet seismic shift in Darwin's thinking. At the time, this was a radical idea. Was he right? Absolutely.

For almost half a century, Peter and Rosemary Grant of Princeton University have been closely monitoring populations of medium ground finches, *Geospiza fortis*, on the Galápagos island of Daphne Major, making annual trips to study the birds' physical traits – weight, body size, beak size and shape – and gathering data on their primary food sources, mating behaviours, and birth and death rates. What they have found further supports Darwin's theory that natural selection brought about by environmental pressures can lead to the development of new species. It also shows, among other things, just how well seeds can influence the evolution of animals. It is in seeds that we can see those 'different ends' Darwin wondered about.

Mitochondrial DNA dating indicates that the 14 species of 'Darwin's finches' in the Galápagos, and one species that made it as far as Cocos Island (Isla del Coco) to the north, all evolved from a single common ancestor within the past 1.5 million years. Also, they're not really finches. That common ancestor, which so ambitiously flew to the Galápagos from the South American mainland all those years ago, appears to have belonged to a family of small 'finch-like' birds called tanagers. That bird's descendants then moved from island to island, adapting to a variety of habitats and food sources – and the arrival of new islands. Geologically speaking, the Galápagos islands are still babies. They only began to form less than 5 million years ago and they're still in the process of forming today. Consequently, the older and the newer islands present vastly different habitats.

Moreover, the Galápagos's location, situated precisely at the equator on the eastern edge of the Pacific, adds further evolutionary pressures in the form of weather extremes. The archipelago is exquisitely vulnerable to the El Niño Southern Oscillation. El Niño years bring heavy rain falls to the Galápagos while La Niña often brings protracted drought.

Medium ground finches are largely seed eaters and have variations in beak size, just as variations in height and body shape exist among humans. Those with beaks at the smaller end of the spectrum tend to subsist on small, soft seeds, many of which are deposited by wind on the rocky landscape. The finches hop all over the terrain to find them, their smaller, pointed beaks suited to exploring all the nooks and crannies. The medium ground finches with larger beaks don't bother as much with small seeds, but they have no problem munching on the seeds of *Tribulus cistoides*, contained within a large, hard, spiny casing called a mericarp. Technically it's a fruit, but it looks like pain for lunch. From the plant's point of view, the hard casing and sharp spines are meant to dissuade seed predation, which works for most everything except the large-beaked *G. fortis*, which breaks the seeds open with a forceful crushing action. In their first decade of observing finches on Daphne Major, the Grants witnessed just how important this beak–seed relationship was.

For years, medium ground finches with small beaks had done quite well scouring the island for small seeds, but that all changed in 1977 with the arrival of a La Niña event. The resulting drought was merciless. No rain fell on the Galápagos for eighteen months.

Many of the plants died, especially those that produced small seeds. The only remaining seeds available were those produced by the *T. cistoides* plant, which is drought-tolerant, but *G. fortis* with small beaks couldn't tackle the large, spiny seeds. As a result, around 80 per cent of the medium ground finches died during that drought. Very few smaller-beaked birds endured, while those finches that survived tended to have large, strong beaks which enabled them to access the tougher seeds. These birds then passed their genes on to the next generation. The Grants found that the average beak size of *G. fortis* on Daphne Major was significantly larger after the drought than it had been before.

'This was an evolutionary response to a natural selection event for the reason that it's the large birds that could eat the large, hard *Tribulus* seeds,' explained Rosemary Grant in a lecture on the evolution of Darwin's finches that she presented at Cornell University in 2018. Grant went on to recount that, a few years after that fateful La Niña, it all changed again. In late 1982, an El Niño event arrived in the Galápagos, and the archipelago spent much of 1983 drenched by heavy, prolonged rainfalls. Over a metre of rain fell that year, estimated to have been the most severe rainfall event in the region in 400 years. The effect on the vegetation on Daphne Major was profound. Diminutive seed-producing plants grew in abundance, yielding large crops of small, soft seeds. Meanwhile, *T. cistoides* plants were completely overrun, smothered by grasses, herbaceous plants and wet weather–loving vines. 'It completely changed the island from a large, hard seed producer to a small seed producer,' said Grant.

The wet season is usually when the finches breed. Indeed, the first heavy rainfall of the season seems to make these little birds enthusiastically libidinous. In 1983 they bred like crazy. What's more, there were now plenty of plants producing small, soft seeds. 'When the drought came two years later, this time it was the smaller birds who had the selective advantage and it was them who survived,' explained Rosemary Grant. By impacting the production of certain seed shapes and sizes, weather events were providing evolutionary nudges in this isolated population of birds. And not over millions of years but twice in less than a decade. The Grants were seeing evolution by natural selection take place in real time.

So it is that seeds changed animals and animals changed seeds, and the evidence is all around us, in ants and squirrels, and in finches that aren't really finches. Many animals are generalists, subsisting on a variety of plants and seeds, but sometimes the pairings are so exquisitely specific that the abundance or absence of a particular plant or fruit or seed is enough to literally change the shape of a species and profoundly alter the course of its population. Such close relationships can have a big impact on plants, too.

Andrew Rozefelds is fascinated by the diversity of life and myriad strategies for survival, especially in rainforests, but he worries about the loss of today's key seed dispersers. He talks to me about cassowaries, the large, long-legged, flightless birds that belong to the same family as ostriches and emus. They reach heights of 1.8 metres, weigh upwards of 70 kilograms, have a tough keratinous protuberance atop their heads, and enormous scaly claws, and look very much like something that belongs on the set of *Jurassic World*. Of course, all modern birds descend from a lineage of avian dinosaurs that survived the mass extinction at the end of the Cretaceous. Indeed, there is a theory that the ability to peck on seeds is precisely what enabled them to survive the prolonged 'impact winter' that brought the Cretaceous to its ultimate end. But although all birds are considered living dinosaurs, cassowaries seem to bear the most striking family resemblance to their Mesozoic forebears.

'Cassowaries are cool!' says Rozefelds with a smile, and he's not just talking about their appearance. He explains that cassowaries' role in maintaining rainforest ecosystems is deeply underappreciated. They're prolific seed dispersers, consuming the fruits and defecating the seeds of more than 230 plant species. Through their role as rainforest gardeners, they operate as keystone species – species upon which many other species, both plant and animal, survive. Yet their populations are declining. In Australia, for example, cassowaries once roamed almost the entirety of tropical northern Queensland, but today there are only 4000 left in three isolated populations, thanks to increasingly fragmented habitats. With isolation come genetic bottlenecks and a higher risk of disease. At the moment, though, they mostly succumb to dog attacks, car strikes, and egg predation by feral pigs. This is not good, says Rozefelds, not for the cassowaries and not for the rainforests they quietly engineer. 'If we lost cassowaries, I suspect there's a good chance it would impact everything.'

Andrea Blackburn has similar worries about the orangutans on Borneo and nearby Sumatra, which are keystone species in their ecosystems and also critically endangered. The orangutans on these two islands are the only ones left in the wild, she tells me, adding that they face 'all sorts of threats – deforestation, fires, hunting and poaching'. By and large, though, the biggest problem is habitat loss, she says.

Orangutans are not the only primates we need to learn more about, especially in terms of their gardening capabilities. Primates, after all, make up a substantial amount of frugivore activity in the tropics. Seed-dispersal activity has been observed in chimpanzees, gibbons, macaques, bonobos, baboons and more. Western lowland gorillas (*Gorilla gorilla gorilla*) in the Central African country of Gabon disperse the seeds of at least 117 different plants. In the dense primary forests of Colombia, brown spider monkeys disperse huge numbers of seeds from a variety of species and in so doing engineer ecosystems in which many other species – from jaguars to sloths – can live. On rare occasions, brown spider monkeys with white fur have been spotted in those forests, and it isn't a good sign.

Dubbed 'ghost monkeys', their pale appearance isn't due to albinism but rather is a genetic glitch that likely signifies inbreeding. This can happen when a population dwindles to the edge of vanishing. There are now so few brown spider monkeys left in the wild, they are classed as of the most endangered primates on the planet. We already have enough ghosts of evolution. We don't need new ones.

There is one primate with a very different story, though. More than once in its evolutionary journey, it was in danger of flickering out of existence. But that was a long time ago, and it has since proliferated to such an extent that it impacts every ecosystem on the planet. That primate is *Homo sapiens*. Humans. Us. As it turns out, the way we've interacted with seeds has quite a lot to do with how the world got to be the way it is. What we do with seeds from here on out will profoundly affect far more than you may realise.

✱ *Why facts have ruffled feathers in the birding world*, p. **238**
   Tawny frogmouths, p. **282**

'GUT-WRENCHING AND INFURIATING': WHY AUSTRALIA IS THE WORLD LEADER IN MAMMAL EXTINCTIONS, AND WHAT TO DO ABOUT IT

Euan Ritchie

In fewer than 250 years, the ravages of colonisation have eroded the evolutionary splendour forged in this continent's relative isolation. Australia has suffered a horrific demise of arguably the world's most remarkable mammal assemblage, around 87 per cent of which is found nowhere else.

Being an Australian native mammal is perilous. Thirty-eight native mammal species have been driven to extinction since colonisation, and possibly seven subspecies. These include:

Yirratji (northern pig-footed bandicoot)
Parroo (white-footed rabbit-rat)
Kuluwarri (central hare-wallaby)
Yallara (lesser bilby)
Tjooyalpi (lesser stick-nest rat)
Tjawalpa (crescent nail-tailed wallaby)
Yoontoo (short-tailed hopping-mouse)
Walilya (desert bandicoot)
toolache wallaby
thylacine

This makes us the world leader of mammal species extinctions in recent centuries. But this is far from just a historical tragedy.

A further 52 mammal species are classified as either critically endangered or endangered, such as the southern bent-wing bat, which was recently crowned the 2022 Australian Mammal of the Year. Fifty-eight mammal species are classed as vulnerable.

Many once-abundant species, some spread over large expanses of Australia, have greatly diminished and the distributions of their populations have become disjointed. Such mammals include the Mala (rufous hare-wallaby), Yaminon (northern hairy-nosed wombat), Woylie (brush-tailed bettong) and the Numbat.

This means their populations are more susceptible to being wiped out by chance events and changes – such as fires, floods, disease, invasive predators – and genetic issues. The ongoing existence of many species depends greatly upon predator-free fenced sanctuaries and offshore islands.

Without substantial and rapid change, Australia's list of extinct mammal species is almost certain to grow. So what exactly has gone so horribly wrong? What can and should be done to prevent further casualties and turn things around?

Up to two mammal species gone per decade

Australia's post-colonisation mammal extinctions may have begun as early as the 1840s, when it's believed the Noompa and Payi (large-eared and Darling Downs hopping mice, respectively) and the Liverpool Plains striped bandicoot went extinct.

Many extinct species were ground dwellers, and within the so-called 'critical weight range' of between 35 grams and 5.5 kilograms. This means they're especially vulnerable to predation by cats and foxes.

Small macropods (such as bettongs, potoroos and hare wallabies) and rodents have suffered most extinctions – 13 species each, nearly 70 per cent of all Australia's mammal extinctions.

Eight bilby and bandicoot species and three bat species are also extinct, making up 21 per cent and 8 per cent of extinctions, respectively.

The most recent fatalities are thought to be the Christmas Island pipistrelle and Bramble Cay melomys: the last known record for both species was 2009. The Bramble Cay melomys is perhaps the first mammal species driven to extinction by climate change.

Overall, research estimates that since 1788, about one to two land-based mammal species have been driven to extinction each decade.

When mammals re-emerge

It's hard to be certain about the timing of extinction events and, in some cases, even if they're *actually* extinct.

For example, Ngilkat (Gilbert's potoroo), the mountain pygmy possum, Antina (the central rock rat), and Leadbeater's possum were once thought extinct, but were eventually rediscovered. Such species are often called Lazarus species.

Our confidence in determining whether a species is extinct largely depends on how extensively and for how long we've searched for evidence of their persistence or absence.

Modern approaches to wildlife survey such as camera traps, audio recorders, conservation dogs and environmental DNA make the task of searching much easier than it once was. But sadly, ongoing examination and analysis of museum specimens also means that we're still discovering species not known to Western science and that tragically are already extinct.

What's driving their demise?

Following colonisation, Australia's landscapes have suffered extensive, severe, sustained and often compounding blows. These include:
- widespread habitat modification and destruction

- the introduction of invasive predators, such as feral cats, red foxes and herbivores (European rabbits, feral horses, goats, deer, water buffalo, donkeys)
- toxic 'prey' (cane toads)
- intense livestock grazing
- changed fire patterns associated with the forced displacement of First Nations peoples and cultural practices
- climate change
- hunting
- disease.

And importantly, the ongoing persecution of Australia's largest land-based predator: the dingo. In some circumstances, dingoes may help reduce the activity and abundance of large herbivores and invasive predators. But in others, they may threaten native species with small and restricted distributions.

Through widespread land clearing, urbanisation, livestock grazing and fire, some habitats have been obliterated and others dramatically altered and reduced, often resulting in less diverse and more open vegetation. Such simplified habitats can be fertile hunting grounds for red foxes and feral cats to find and kill native mammals.

To make matters worse, European rabbits compete with native mammals for food and space. Their grazing reduces vegetation and cover, endangering many native plant species in the process. And they are prey to cats and foxes, sustaining their populations.

While cats and foxes, fire, and habitat modification and destruction are often cited as key threats to native mammals, it's important to recognise how these threats and others may interact. They must be managed together accordingly.

For instance, reducing both overgrazing and preventing frequent, large and intense fires may help maintain vegetation cover and complexity. In turn, this will make it harder for invasive predators to hunt native prey.

What must change?

Above all else, we genuinely need to care about what's transpiring, and to act swiftly and substantially to prevent further damage.

As a mammalogist of some 30 years, the continuing demise of Australia's mammals is gut-wrenching and infuriating. We have the expertise and solutions at hand, but the frequent warnings and calls for change continue to be met with mediocre responses. At other times, a seemingly apathetic shrug of shoulders.

So many species are now gone, probably forever, but so many more are hurtling down the extinction highway because of sheer and utter neglect.

Encouragingly, when we care for and invest in species, we can turn things around. Increasing numbers of Numbats, Yaminon and eastern barred bandicoots provide three celebrated examples.

Improving the prognosis for mammals is eminently achievable but conditional on political will. Broadly speaking, we must:

- minimise or remove their key threats
- align policies (such as energy sources, resource use and biodiversity conservation)
- strengthen and enforce environmental laws
- listen to, learn from and work with First Nations peoples as part of healing Country
- invest what's actually required – billions, not breadcrumbs.

The recently announced Threatened Species Action Plan sets an ambitious objective of preventing new extinctions. Of the 110 species considered a 'priority' to save, 21 are mammals. The plan, however, is not fit for purpose and is highly unlikely to succeed.

Political commitments appear wafer-thin when the same politicians continue to approve the destruction of the homes critically endangered species depend upon. What's more, greenhouse gas emissions reduction targets are far below what climate scientists say are essential and extremely urgent.

There's simply no time for platitudes and further dithering. Australia's remaining mammals deserve far better. They deserve secure futures.

✱ *In the shadow of the fence*, p. **136**
   Noiseless messengers, p. **162**

ISOLATION

Paul Biegler

On a cool night in the suburbs of Los Angeles in December 1978, Jack Morris was at a house party. It was one o'clock in the morning and things were getting tense. Morris was 18 years old and hadn't long been released from juvenile detention.

'We were all drinking,' he remembers. 'Two guys went out into the back yard to fight, that fight ensued, and then the guy that was fighting said he didn't want to fight no more, and he ran into me, and he ran into another guy.'

Morris is seated at his desk at the St John's Community Health Clinic in a squat brick building south of downtown LA, where he helps people who've left prison get jobs, housing, manage drug problems and avoid re-entering the prison system. He's spruce, with thick greying hair cut to a medium crew and he's wearing a dark shirt, and smart, tinted reading glasses. Morris speaks confidently, but when he recalls what happened, his voice fades.

'And then I pulled out a knife and stabbed him. He died right there.'

Morris graduated to adult prison. He went to San Quentin, Corcoran then Tehachapi, doing solitary confinement in each. But out in the general population there was human contact. A handshake, a conversation, a gaze being met. In August 1991, however, all that changed. Morris was transferred to Pelican Bay, a supermax prison in northern California where the inmates were housed in windowless, poured concrete cells that measured 2.4 x 3 metres, around half the size of a standard car parking space.

'When the door is closed, even though it is perforated plate steel, you could feel your soul being sucked out,' he recalls. 'You're standing there naked with what they handed you and you're just looking at the cell and you realise there is nothing in here but a concrete box, stainless steel sink, a toilet. You're standing there ... barefoot on cold cement, and you're saying, "Well, this is it".'

What Morris didn't realise was that a sentence to solitary came with hidden extras. Soon, like hundreds of fellow prisoners, he would get high blood pressure. His risk of heart disease would go up. He would get symptoms of anxiety, sleeplessness and depression. His knees and back would ache and he'd be taking anti-inflammatory medication. But few people cared about the plight of hardened criminals doing time.

Then the pandemic hit, and much of the world was forced into its own brand of lockdown isolation. Rates of depression and anxiety skyrocketed. The American Academy of Pediatrics declared a national emergency in children's mental health. Now everyone was interested in the very same question: what does isolation do to bodies and minds? Scientists tackling that question have come up with some disturbing answers. Isolation, it seems, changes us down to our core, right down to the level of our genes.

Blue genes?

These days Steve Cole is a professor of medicine at the University of California, Los Angeles, but in the late nineties he was a virologist studying HIV. And he was perplexed. Something odd was happening with a subset of the gay men who were getting the virus. Their immune system was declining faster, and they were dying earlier.

'The ones that were most vulnerable were those who were most socially marginalised, especially the guys who were in the closet,' Cole recalls. 'How was it that the HIV genome operated differently in the body of a person who was living in the closet, and constantly afraid of being discovered and losing his job, or his family, or his friends?'

Cole embarked on a series of studies to find out, and unearthed something astonishing. Gay men with a 'socially inhibited' temperament were switching on their 'fight and flight' sympathetic nervous system. They were pumping out stress hormones, including one called noradrenalin, which binds to a white blood cell that's key in the fight against HIV, the T lymphocyte. Those lymphocytes were then winding back production of proteins called interferons, critical defenders against viruses.

'There was a set of programs that were built into the human immune system, that caused threatened people to throttle back on their antiviral immune responses,' says Cole.

Incredibly, the social truth of sexual orientation was playing out at the level of these men's infection-fighting cells. Cole presented his findings at a think tank and, afterwards, was approached by a man with an unusual request. His name was John Cacioppo and he was a pioneer in the scientific study of loneliness, whose links to physical diseases such as heart disease, Alzheimer's and high blood pressure were only just being teased out.

Cacioppo wanted to know if the stress of social isolation could do similar things to the immune system and might somehow be elevating the risk of chronic disease. Cole jumped at the chance to work with him.

'We did this initial study. It was painfully small, like, 14 people. A really small number of chronically lonely individuals, and another complementarily small number of chronically socially integrated individuals. And we just took some white blood cells that John had stored in freezers and did this RNA profile on them,' Cole remembers.

That profile homed in on the activity of genes in the white blood cells. It was a small study, but the results were emphatic.

'If you look at the genes that were most reduced in activity in lonely people, a surprisingly high proportion of them were the genes that were involved in type one interferon responses, these basic, innate antiviral responses,' says Cole.

'That was really stomped down in these lonely people in a way that looked a lot like the gay men in the closet.'

The stress of loneliness – the subjective, unpleasant experience of social isolation – was doing similar things to the stress of hiding one's sexuality. Something, however, didn't gel. How could a stunted response to viruses put lonely people at risk of illnesses such as coronary heart disease, that seemingly had nothing to do with a virus?

Cole's study had picked up a clue. If lonely people were making less interferon, genes linked to production of something else were in overdrive.

Lonely people were churning out more proteins, known as cytokines, that ramp up inflammation. And inflammation is central to heart disease and increasingly thought to play a role in cancer and Alzheimer's, the very diseases Cacioppo was interested in.

'It was pretty clear that there was this teeter totter regulation where the human immune system, which is generally pretty heavy on antiviral capacity and kind of stomps down on inflammation, was recalibrating in the opposite direction in these lonely people,' says Cole.

According to Cole there are sound evolutionary reasons the body would do this. Our default mode is to be prepared for the infectious threat we see all the time – socially transmissible viruses. A rarer event is a wound which, if it gets infected with bacteria, can kill us quickly. It is the inflammatory response, and the mobilisation of different types of white cells, such as monocytes, that mop up these bacterial invaders.

The upshot, says Cole, is that we need to be able to pivot, rapidly, from an antiviral to an antibacterial footing, something the stress response achieves well.

'Loneliness is just a surprisingly good way of being threatened. The primary safety signal for most humans is being a member of some kind of supportive community, a family, a tribe.' The body, then, sees loneliness as a cue that injury, and bacterial infection, are just one predator away.

When the pandemic began, Cole was a go-to guy for the media, who were desperate to uncover what the science showed about millions of people in enforced isolation. Cole realised he could only make inferences from previous studies. Then he realised he might have the answer right in front of him.

For many years Cole has worked closely with primate researcher John Capitanio, a psychologist at the University of California, Davis. Around 2016 the pair, in a team that included John Cacioppo, began a study of rhesus macaque monkeys at the California National Primate Research Centre. It's a sprawling facility on the outskirts of Sacramento, where scores of monkeys cohabit in half-acre field cages equipped with play gyms, mini-Ferris wheels and tree branches.

The team wanted to catalogue the behaviour of lonely and non-lonely monkeys during two weeks isolation in a smaller cage. In 2020, when the media queries started pouring in, Cole realised the experiment was a near-perfect model to study the effects of lockdown. So the team switched tack. What would the mere fact of being separated from its friends do to the monkey's immune system, they wondered? A lot, as it happened.

The results, published in *Proceedings of the National Academy of Sciences* in 2021, found the monkeys dialled up, by around 9 per cent, production of white blood cells, called classical monocytes, responsible for inflammation. Cells that make interferon and fight viruses, such as lymphocytes, plummeted by up to 50 per cent. How soon did it all start? Within 48 hours of the animals entering the cage. It was a dramatic demonstration of the biological force of isolation.

Not everyone, of course, experiences lockdown alone. Parents, for example, have to care for children. The researchers knew older primates are predisposed to care for juveniles, so they tried something else. A month after each monkey was released they put it back into isolation, this time caged with a young monkey.

'That made a huge difference in terms of how the adult monkey's biology responded to lockdown,' says Cole. 'The monkey that goes into the same setting, with an infant there, doesn't show any material ramp up in inflammatory cell, monocyte production, and doesn't show any material ramp down in antiviral biology.'

Put a monkey in lockdown with caregiving, they had found, and all those stress-related changes in immunity seem to melt away. The reason why the mere presence of a needy child could taper the adult monkey's stress response may well come from a curious study that Cole and colleagues published in 2013.

They analysed gene expression in the white blood cells of 80 healthy adults and, at the same time, asked them how, on balance, their lives were going. Some people scored higher on scales that measure a sense of purpose and meaning in life, so-called 'eudaimonic' wellbeing (in the writings of Aristotle, *eudaimonia* refers to 'human flourishing'). Those people, the team found, were turning down the threat-related immune response.

Cole has since gathered evidence that being connected to a deeper purpose activates brain areas linked to motivation and 'wanting', including the ventral striatum and nucleus accumbens. Those areas can quell the effects of the brain's anxiety centre, the amygdala, whose job is to put the body on a threat footing.

'When those monkeys engage in caregiving, and when humans engage in this kind of virtuous and striving and generative activity, what you're getting is a lot of activity in this seeking, hoping, wanting section of your brain that essentially vetoes some degree of threat-related biology,' says Cole.

It is a facility, Cole believes, that likely arose to push parents to face danger for the sake of their children. If a parent is to run inside a burning building and save a child, for example, they must put the risk of self-harm to one side.

Closing time

No animal study maps perfectly onto the human condition. But it's hard to avoid speculating that aches in Jack Morris's back and knees were manifesting a change in his inflammatory biology.

Inflammation has also been linked to high blood pressure, but it's likely the stress of isolation, and a rise in stress hormones like cortisol, contributed too.

Morris also seemed driven to find a higher purpose. He studied maths. He took legal training and helped fellow inmates lodge writs. He read thousands of books on everything from quantum physics to the migration patterns of the peregrine falcon.

'You have to figure out a way to exist. You can't sit there and stare at the wall because then the wall is going to start moving on you. I didn't know how to draw. I learned it, because it gave me something to do eight to ten hours of that day.'

There were pencils, but no colour. Morris drew a bird, with intricate feathers and strange, cord-like appendages in red, blue and green. He made the colours from Skittles and M&M's. Somehow, in that prison within a prison, Morris was impelled to find meaning. But something else was happening too. The hunger for human contact, so familiar to people who've spent time alone, was ebbing away. Morris could yell to the eight other people in his 'pod' through the dime-shaped holes in the steel door. But, as the years passed, he stopped.

'You don't even want to communicate no more. Your world becomes smaller and smaller and smaller. Somebody calls up to you saying, "What are you doing? I haven't heard from you in two days." And we're in the same pod. And my cell is next door to him.'

What is it about isolation that turns an innately social being into one that no longer seeks the company of others?

In pursuit of happiness

A bit over a decade ago Gillian Matthews was elbow deep in her PhD at Imperial College London. She was researching addiction – specifically, the effects of cocaine on an enigmatic structure in the mouse brain called the dorsal raphe nucleus or DRN. But when you inject cocaine into mice, things can get, well, interesting.

'Typically the mouse on cocaine is quite excitable,' says Matthews. 'You'll put them in a separate cage overnight for 24 hours, so that they don't interrupt their cage mates.'

Matthews had a control group of mice getting salt water injections, so, to keep conditions equal, they were also isolated for 24 hours. When Matthews examined the mouse brains afterwards she found a major change in the cocaine group – connections in neurons that feed into the DRN, and drive its activity, had grown way stronger. But something weird was going on. The mice that got salt water had an almost identical muscling up of those same neurons.

'My initial reaction was that I'd probably mixed up the syringes, given the saline mice the cocaine,' Matthews remembers. She repeated the experiment. Same result. The effect was real. The simple fact of isolation looked to be pumping up neurons that drive activity in the DRN. But why?

Until now, Matthews had been recording DRN activity in brain slices from euthanised mice, whose neurons can remain active for 12–14 hours. To get a better handle on the DRN meant studying live mice using different techniques. So, in 2013, Matthews crossed the Atlantic to take up a post-doc role in the lab of Professor Kay Tye, then at MIT.

They began experiments in animals bred so that fluorescent indicators could track the influx of calcium into their neurons, which happens when the neurons are firing. It's called calcium imaging – brighter neurons mean more activity. For 24 hours the mice were either held in isolation or caged with a bunch of other mice. When the time was up they had each mouse say hi to an unfamiliar young mouse.

The mice that had been group housed had a bit more activity in the DRN neurons. But in the mice that had done solo time, it was something else again.

'These neurons go crazy,' said Tye, shortly after the experiment's publication in 2016.

DRN neurons looked to be tracking the social state of the mouse, firing up in isolated animals when they finally reconnected with others. But could those brain cells be doing even more? Were they actually driving behaviour? To find out Matthews and Tye turned to optogenetics, a technique that introduces light-sensitive proteins into neurons, which can then be activated by a given wavelength of light. The team engineered things so the DRN could, effectively, be turned on at the flick of a switch. They put the mouse in a cage with a mesh wall at one end, behind which another mouse was going about its business. And flicked the DRN switch. The effect was immediate, and dramatic.

'We found that they increased the amount of time that they wanted to spend with another mouse,' says Matthews. 'They increased their social preference, so they spent more time on that social side of the chamber.' It was a stunning finding. The DRN was, almost single-handedly, deciding whether mice would be shrinking violets, or party animals.

Matthews had another tool at her disposal. Optogenetics can also silence neurons. When mice go into isolation for short periods they get convivial on release, as if to make up for the social deficit. The team put mice into lockdown for 24 hours, and then used optogenetics to turn off the DRN neurons. 'When we silenced these neurons after animals had gone through 24 hours of isolation, we found that it suppressed that rebound social contact,' says Matthews.

The DRN, they had discovered, was crucial to the process of reconnecting after brief isolation. Matthews, Tye and other researchers at the Salk Institute in California, where Tye is now based, have set out what they think happens over longer periods of isolation when many people become socially withdrawn.

In two key papers, published in 2019 and 2021, they describe the DRN as part of a system that works like a thermostat. Reduce your contacts and the system will prompt you to mingle and get your social 'temperature' back to the set point. Matthews and Tye call this 'social homeostasis'.

But deprive someone of social contact for months on end and they can get antisocial, content with the new normal of fewer entanglements. Matthews and Tye think this is a 'set point adaptation', a reset of the thermostat that winds down our social needs. You might feel fine in the trackie daks and slippers of lockdown, but emerge into your former social milieu and things get uncomfortable. It now feels overcrowded and people can react with avoidance, aggression or social anxiety. So what should people do to keep things steady in lockdown, and to re-adapt when they come out?

Remote control: The Zoom effect

Dr Louise Hawkley is a principal research scientist at NORC at the University of Chicago and has studied loneliness for more than two decades. When the pandemic hit, Hawkley was on a team that had been surveying a group of older adults as part of the National Social Life, Health, and Aging Project.

'There was all this hype around video calls,' says Hawkley. 'That's your answer, you better learn how to do video calls because then you're going to get your best quality interaction with other people.'

Hawkley convened a new team to survey around 2500 people from the cohort, all aged 55 and up. They asked respondents to rate their levels of happiness, depression and loneliness, which could be compared with ratings from the same group taken in 2015, pre-pandemic.

People also reported their level of contact with family and friends, how much was in-person, how much was via video and phone calls, email and messaging, and how it all tracked over the pandemic. Hawkley's findings, published in the *Journal of the American Geriatrics Society* in 2021, showed that around 40 per cent of people had reduced in-person contact.

The effects, too, are unsurprising. 'The more you stayed at home and didn't have contact, the more lonely you became, the more depressed symptoms you had, the less happy you were,' says Hawkley.

Most worrying, though, was the impact of remote contact. 'Could they improve that by having more virtual interactions? The take-home seems to be "no",' she says. Hawkley's team found that between 16 per cent and 26 per cent of respondents increased their levels of remote contact during the pandemic.

'It seemed to make no difference,' says Hawkley. 'It did not alleviate loneliness to any significant degree. It did not alleviate depressive symptoms or improve happiness.'

There is little data on why this is the case, so Hawkley can only offer her intuition, albeit drawn from decades of research.

'It's the whole sensory experience. It's not just what we see and hear but all the subtle cues we get from, call it body odour, pheromones, whatever ... And there are subtleties in body posture, physiognomy, that essentially change with emotions. That is very hard to detect accurately on a virtual platform.

'We all have that need for assurance, for security, and we don't give it to babies by having them watch a screen. It requires this very immediate detection of signals, eye to eye, hand to hand, chest to chest, just to round us out as what we are as people. Because we aren't islands. We are not sufficient for ourselves.'

Hearts and minds

In December 2016 Jack Morris was told, after 30 years of continuous solitary confinement, that he would be transferred to general population in a prison downstate.

'I spent Christmas staring out of a window. Out of a window. And I saw snowflakes falling from the sky and that's after living in a cell for decades and not seeing no night sky, or no sun, or no clouds,' he says. 'For weeks I had headaches. I was seeing colours. I'd see the colour of dirt. I'd see the colour of grass. Because in there it's strictly concrete, you don't see colour. You walk on concrete, you sleep on concrete, you eat on concrete. You stare at concrete.'

Reconnecting with people came with a distinct set of challenges.

'I didn't know how to socialise. How can I go up to somebody and talk to them? What am I going to do, look at 'em and talk to 'em? Shake a hand? You don't do that. That's foreign to you. It's no longer something you know about. It's something you have to learn all over again.'

The process wasn't helped by the profound anxiety Morris felt being around people again. 'Your heart, you could literally hear it pounding in your chest. Somebody would just run by. You're turning around, you're jumpy, you're jittery. You can't sit down. You can't get any relaxation.'

Released from prison in 2017, Morris has slowly inched his way back into society. Millions of people around the world are recovering from their own, less extreme struggles with isolation.

'You do need to be conscious that there are long-term changes that could have impacted how long it will take people to change their behaviour and to feel comfortable and to feel less anxious,' says Matthews.

But deep in all that neuroscience there is also a message of hope. 'If we have learnt anything about the brain it is that it's really good at adapting. I think people will adapt back, and I think the brain will adapt too.'

✱ *The psychedelic remedy for chronic pain*, p. **195**
Do we understand the brain yet?, p. **262**

WHY FACTS HAVE RUFFLED FEATHERS IN THE BIRDING WORLD

Tabitha Carvan

For such a small, hapless bird, the Night Parrot sure gets humans worked up.

It's the holy grail of ornithology. It's legendary, fugitive, mysterious, almost mythical: the ghost bird; the Thylacine of the air. To see one would be like 'finding Elvis flipping burgers in an outback roadhouse', 'a cockatoo in a snowstorm', or, in more straightforward terms, 'a giant gold nugget'.

For more than a century, the Night Parrot has inspired not only colourful language, but obsessive odysseys, all fruitless, eluding everyone it lured into its desert habitat.

Until 2013, when the seeming impossible happened: someone actually got photographs of this 'white whale of the bird-watching world'. It was a moment described by the *New York Times* as 'one of the greatest stories of species rediscovery in recent times'.

And it *was* a great story! In conservation news, it made for a nice change to have a happy, hopeful headline instead of a death notice. There was a great hero too, cast in the leading role.

You might think the hero would be the bird, but one of the few things we know about Night Parrots is they don't like attention. In the photographs, the green and black speckled bird appears to be cringing under the spotlight. And quest stories are never really about the object being sought, anyway.

No, the hero was the quester: John Young, the naturalist who took the photo. Described as a 'bush detective', Young had spent most of his life looking for – and often finding – rare birds in remote Australia, and he had the stories to match.

In his gung-ho determination to find the Night Parrot, he told the ABC he was 'like a terrier dog': 'I couldn't leave it alone, it was like it was blood in my veins and I just didn't want to stop.'

In the dozens of articles published around the world about the bird's rediscovery, Young looks exactly the part of the larrikin hero: Crocodile Dundee meets Steve Irwin, with Merv Hughes's moustache.

More articles would soon follow. While employed as Australia's leading Night Parrot expert by the Australian Wildlife Conservancy, Young identified more and more locations for Night Parrot populations over the coming years, photographing their nests, finding feathers, and recording calls.

It was an onslaught of unbelievably good news, not to mention amazing PR for the Australian Wildlife Conservancy. For this critically endangered bird, John Young really was the hero we all needed.

'It's a hugely interesting story,' Dr Penny Olsen says of the history of the Night Parrot. 'And these characters get involved because there's kudos if you've seen one or found one, so that leads to some bad behaviour, you could call it.'

Dr Olsen is a different kind of Night Parrot expert. She is the author of the definitive book on the subject, *Night Parrot: Australia's Most Elusive Bird*, and Honorary Professor at the ANU Research School of Biology.

She has been on her own version of a quest for the Night Parrot, only her quest didn't take her into the outback, but deep into the stacks of the National Library of Australia. 'I look for everything I can find. Absolutely everything, going back historically.'

Dr Olsen has a holy grail too: 'I like the truth, as much as you can know it. Science. Fact. Whatever. I like to set the record straight.' She believes the public needs to have faith in facts, now more than ever.

Before even starting research on her Night Parrot book, Dr Olsen had suspicions that Young's celebrated findings might not be all that they seemed. At the library, she dug deeper and deeper until she had accumulated 'a mountain of material' supporting her theories. Then, in her book, she set the record straight.

In what *The Australian* called 'a stinging attack', Dr Olsen suggested the bird which featured in Young's original photographs was injured and had been staged for the photoshoot, not found in the way he claimed.

She also noted inconsistencies and question marks surrounding some of Young's subsequent findings, including nests, eggs and feathers, saying they should not be taken as definitive proof of Night Parrot populations.

'To some extent, I was defending my profession,' she says of what she wrote. 'I think science should be as close to the truth as you can get.'

Young's data was being used to inform decisions about funding and planning for Night Parrot conversation. The bird might feel mythical, but it lives in the real world, and the consequences of misrepresenting its population size and distribution were serious.

After her book was published, Dr Olsen says she received a death threat. Her critics emailed booksellers demanding they remove 'that terrible woman's book' from the shelves. She was accused of ruining Young's reputation. Many people in the birding world believed she had ruined her own reputation too.

It was a stressful time, she says, but she was confident in her research. She had gone over absolutely everything.

'There's a side of her that's a little bit of a terrier,' Professor Robert Heinsohn, from the Fenner School of Environment and Society, says of Dr Olsen. 'You have to have that attitude to keep going on research like hers. She knew she was on to something.'

Not too long after the publication of Dr Olsen's book, an

uncropped version of Young's celebrated Night Parrot photo came to light. In the photo, you can see wire mesh in the background, leading to concerns that the bird might have been illegally captured, as Dr Olsen had suggested.

Young resigned from the Australian Wildlife Conservancy, which went on to investigate his fieldwork, including those results highlighted by Dr Olsen. A panel subsequently found that she was right to have doubts: three purported Night Parrot nests could not be proven to be such, and one contained fake eggs; a field recording of a Night Parrot call was actually a publicly available recording of a call from a different location; and there were discrepancies regarding where Young said he had collected a Night Parrot feather.

Summarising the panel findings, *Audubon* magazine wrote that Young 'may have fabricated just about everything he reported about new populations and nesting sites of the birds over the past two years'.

There was no great media fanfare vindicating Dr Olsen. There were no headlines calling her 'the library detective'. She says she received one apology from a former detractor, but that was it.

'Penny's integrity is as pure as can be, but she has worn the cost of calling out the misinformation,' says Professor Heinsohn. 'It's unfair, but it happened.'

When Dr Olsen graduated from ANU with an honours degree in science, it was at a time when fewer than 20 per cent of graduates of such degrees were women. After university, she was hired by the CSIRO – starting at a lower grade than her male peers – and was, at the time, the only female research scientist in the Division of Wildlife Research. She was discouraged from doing fieldwork in the belief it wouldn't be safe for her.

She went on to do her PhD, with her thesis being the first Australian study to prove that DDT thinned raptor eggshells.

'She really got raptor research going not just in Australia, but worldwide,' says Professor Heinsohn. She went on to publish more than 130 papers and chapters on ornithology.

'She was a fantastic scientist,' Professor Heinsohn says. 'She still is a fantastic scientist, but now she's pivoted more into science communication.'

Dr Olsen has written 30 general interest books on ecology, focusing mostly on birds, a body of work which has been recognised with an Order of Australia. She has been awarded the Royal Zoological Society of New South Wales Whitley Award for her books a record-breaking seven times.

'The whole progression of her life is just amazing,' Professor Heinsohn says.

But if you ask Dr Olsen about her career, she answers, 'I've always been a bit surprised to find myself places. I'm not a strategist in that way. I've bumbled through one way or another. I met some fabulous people. I've been lucky ... I don't know. Sorry, that's a bit of a rambling answer.'

It's not a great answer, no. But it's a true one, and sometimes that's more important. In science, surely, it's more important.

'I view her as Australia's preeminent ornithologist,' Professor Heinsohn continues. 'She has dedicated her life to it, and she is sitting there at the pinnacle of what we have in this country. She should be considered to be like that by everybody.'

But we give the airtime to those who fit the bill for the story we've already written in our mind.

In the comment section of that *New York Times* article about the rediscovery of the Night Parrot, one commenter, Bart from Los Angeles, notes: 'I'd watch a documentary on John Young. The part about him is better than fiction.' The article doesn't mention Penny Olsen.

'Interestingly, women hardly figure among the searchers and sighters,' Dr Olsen notes in her book's introduction. Since the quest for the bird began, it's been one tale after another of white men heading into the harsh desert landscape, fuelled by bravado and the desire to be The First.

'I don't look the part,' Dr Olsen says. 'I know I haven't always been taken seriously. I'm just a woman who sits on her bum at her office in the university. What would I know?'

She knows a lot. Professor Heinsohn says if he could download the contents of anyone's brain for posterity, he would choose Dr Olsen's.

John Young knows a lot, too. He is an exceptionally skilled naturalist – Dr Olsen would tell you that herself. He dedicated years of his life to finding a Night Parrot, and did so, one way or another. The location of that first bird led Young and ecologist Dr Steve Murphy to find other individuals in the area, contributing enormously to what we know about Night Parrots.

But this isn't really a story about who knows what. It's about who we listen to, and why.

Here's something else about John Young which feels relevant. In 2006, he told the media he'd discovered a new species of bird, the Blue-fronted Fig Parrot. The news elicited a personal congratulations from Queensland's Minister of the Environment.

But when a forensic photography expert analysed Young's photo of the bird, he found it had been photoshopped to change its 'front' – the brow – from red to blue. Young denied the accusation but was also unable to supply the original photo to back himself up. He said he had deleted it. No more was heard about this Blue-fronted Fig Parrot. There are other stories too, ones which might raise your eyebrows. *Audubon* reports on doubts about Young's reliability dating back to 1994.

And yet it was so easy for us to listen to him, and so hard for us to hear that other terrier, Dr Olsen. What does that say about us?

'We're all enablers in some ways,' Dr Olsen says.

We want it to be a good story about a mysterious bird and a heroic 'bushman'. What a drag to have the truth get in the way.

And you can hardly blame the people who write the stories

for capitalising on the human interest angles. When every other headline about species loss is just more of the same, you do what you can to get the reader's attention.

In Queensland, there are only 15 Night Parrots that we know of. The national total is likely to be under 250 individuals, dispersed widely. Their biggest threat comes from fire and feral cats – a shockingly mundane enemy for such a hallowed bird – and habitat loss. Unfortunately for Night Parrots, the ground beneath their spinifex nests is rich with mineral resources.

To protect them, we need evidence-based data, and there are rangers and skilled amateur ornithologists and scientists out there collecting it. And there are even more people working away in offices, on papers and proposals and communication strategies, to raise funds and to lobby for protection measures.

There's no mystique to this work. It's just difficult, and important, and true. In science, such things should demand our attention.

✳ *A mystery of mysteries*, p. **203**
   Gaps in the research, p. **259**
   Tawny frogmouths, p. **282**

DARK SKIES

Karlie Noon and Krystal De Napoli

Imagine that you are lying under unpolluted rural skies for the first time in your life. You have been blessed with ideal observation conditions: the skies are clear of clouds and the area is not affected by light pollution. There is no artificial glow in the sky drowning out the delicate celestial features above you, and the stars that you see are sharp.

Allow yourself the time to look up, take a breath and let your eyes adjust to the darkness encompassing you. Waiting is a crucial step in this observation routine. Our eyes detect light using two light-sensing cells, or photoreceptors, known as cones and rods. Cones operate best in very bright conditions and allow the human eye to detect colours; rods are far more sensitive to light and operate best after we spend at least 10 minutes adapting to the dark – although it takes approximately 20 minutes for our eyes to become fully 'dark adapted'.

The brightest stars are apparent to you immediately – perhaps you recognise the familiar outline of the Southern Cross or catch a glimpse of the transit of a bright planet. Each minute you wait, the more your eyes adjust. Slowly, new stars become visible. Over time, it seems that in every previously dark spot you focus your eyes on, a new star is waiting to be found. The darker the conditions, the greater the quality of your observations. In the right setting, you will find that you are no longer looking only at the clusters of bright spots spattered across the sky, but are beginning to notice the rare regions where there is no light. In one section of the night sky in particular, you may notice that there are pools of no light, framed and illuminated by translucent space clouds that appear to tear the sky in two. You have found the Milky Way, known as the Sky River by many Aboriginal and Torres Strait Islander peoples.

For the Kamilaroi and Euahlayi people, the Sky River is called Warambul, and the dark patches represent flowing water while the surrounding stars are small fires and camps. In a completely different area of Australia, the Yolŋu people believe that when they pass from this world they are taken to Baralku, the spirit-island in the sky where camp fires burn around the great river.

Dark skies and the features within them play a significant role in many Indigenous knowledge systems. For an oral culture with a deeply and intrinsically interconnected knowledge structure, the skies are the melting pot and reference point for much of this information. They are the stage upon which stories unfold. They inform us about the land below and guide us on where we are going, what we need, and who we are, were and will be.

For 65 000+ years, the history of this land has been mapped to features and events in the sky and carried through the generations by the power of language, song and dance. In the time preceding colonisation, there was no need to travel great distances to find a remote area that allowed the experience of the full, unimpeded beauty of the night sky – it was available wherever you found yourself. On a moonless night under pristine dark skies, the human eye may be able to detect anywhere from 2000 to 5000 stars unaided. In stark contrast, due to higher incidences of light pollution caused by unbalanced urban lighting, our modern metropolitan city skies offer visibility of a mere 100 or so stars.

It is not hard to understand that if even the brightest of stars are at risk due to the artificial lightening of our night sky, then fainter features like the Milky Way are entirely vulnerable. The Milky Way is already barely visible in the majority of metropolitan and suburban areas, and even in our rural towns. Other celestial features vulnerable to increased light pollution are the Magellanic clouds, a pair of faint dwarf galaxies that orbit the Milky Way. They are native to Southern Hemisphere skies, appearing as pale smudges of light that never set below the horizon to observers at latitudes of approximately 28 degrees south or more.

If you find yourself in blissful dark-sky conditions and straighten your arm upward, you can cover the larger of the two Magellanic clouds with a closed fist. The width of a human fist approximates a distance of 10 degrees across the sky – this knowledge is a useful tool for Aboriginal and Torres Strait Islander astronomers in measuring celestial distances. To communities in Yirrkala, Arnhem Land, the Magellanic clouds represent two sisters. The older sister resides in the Large Magellanic Cloud (LMC), and the younger sister within the Small Magellanic Cloud (SMC). Yirrkala lies 12 degrees south of the equator and observers there witness the older sister falling below the horizon in the midst of the dry season, leaving her younger sister to fend for herself alone in the sky. When the wet season approaches, the sisters reunite above the horizon, signposting the significant seasonal change the community can expect to experience.

Dark sky constellations

These vulnerable faint areas of the night sky are known by cultural astronomers as dark sky constellations. They are highly valued by Indigenous communities and are a unique way of viewing the sky, quite different to European constellations, which are constructed from groupings of stars. For Indigenous astronomers around the world whose astronomical practices have propagated for millennia prior to the introduction of artificial lighting to the natural environment, dark sky constellations remain a core part of their astronomical traditions.

Dark sky constellations are located in the pools of darkness that form when light from the Milky Way is obscured by gas and dust. As light from distant stars comes in contact with these obstructions, the starlight is absorbed, preventing it from reaching our eyes here on Earth. These dark pools form in several areas of the sky, all well known and understood by Indigenous peoples. Each has its own meaning and purpose to different clan groups across our continent and around the world. However, the features are subtle, requiring not only the darkest skies but also astute observers. As such, they are quickly becoming extinct from view in locations affected by light pollution.

The International Astronomical Union (IAU) – a collective of approximately 12 000 astronomers from 90 countries that promotes and advances astronomy – formally recognises 88 constellations that are familiar shapes made from stars, none of which are dark constellations. However, in areas not yet consumed by light pollution, these dark spaces are still observed and known.

It is within the depths of this darkness that familiar animalistic forms reside. Euahlayi and Kamilaroi astronomers observe Bandarr the Kangaroo snuggling up underneath the Gawarrgay Dark Emu on the banks of Warambul. They are not alone, as the two find themselves in the company of celestial crocodiles whose heads form from the belly of Gawarrgay between September and October. The crocodiles are said to have eaten the wives of Baayina, the creator being, at Coorigal Springs in Lightning Ridge, New South Wales. Baayina freed his wives by killing the crocodiles in a great battle, which in turn created the Narran and Coocoran lakes. Baayina demanded that the crocodiles stay on Earth to protect women's sacred ground and they were forbidden from going to Bulimah, Sky Camp. They can be seen in the waters of the Milky Way, going towards Bulimah but never making it. The story is one of death and resurrection, and signals to the Kamilaroi and Euahlayi peoples that it is time to conduct initiation ceremonies. Interestingly, crocodiles have been extinct in this region for over 40 000 years. Sadly, access to these ancient depictions in our night skies is also now under threat, as discussed later in this chapter.

The animalistic features in the skies have different interpretations depending on which community has developed the knowledge. As these constellations often visually mimic the physical form of their earthly counterparts, similar interpretations of them are shared globally by other Indigenous astronomers. In central Brazil, Tupi astronomers see a very similar bird-shaped dark sky constellation in the Milky Way that overlays our Dark Emu. The shape represents their native bird, the rhea. Coincidentally, the rhea is not only also a large flightless bird, it shares a similar breeding cycle to the emu. Like Gawarrgay, the celestial rhea's head is also made up by the pool of darkness that is the Coalsack Nebula, with the stars of the Southern Cross described as holding the head of this enormous celestial beast in place lest it break free and guzzle up all of the world's water.

These features assist in understanding the behaviour of animals on the ground, learning about seasonal and environmental changes and long-distance navigation, and can also serve as a calendar for ceremonial events. The stars are crucial reference points for knowledge systems relating to astronomy, ecology, medicine, design, history and all things in between. As such, the skies can be thought of as being equivalent to a library for oral cultures, where each star is a book you can call upon to unlock the traditions and knowledge associated with it.

Light pollution and its impacts

Access to this knowledge is becoming increasingly compromised due to the growing impact of light pollution. Light pollution is insidious in that it has become so normalised in modern-day society that it largely goes unnoticed. We have become desensitised to the presence of light around us in unnatural settings. In metropolitan areas, it is accepted that the skies are often illuminated by an artificial glow instead of stars. Western society has become so used to the presence of light that the idea of darkness can provoke fear in many people. In contrast, Aboriginal and Torres Strait Islander peoples have been sleeping under the night sky, enjoying, memorising and living in harmony with its features and cycles since time immemorial. Traditionally, darkness was not feared by Indigenous communities, but valued as offering access to the night sky.

Some of this over-lighting has been driven by legitimate concerns for women's safety in public areas, with the belief that an environment with more lighting is a safer one. Recent research carried out by design company Arup with XYX Lab of Monash University suggests that it is not as simple as increasing the amount of light and that, in fact, overly lit areas feel more unsafe for women as they often lead into low-lit areas, leaving women incapacitated while their eyes adjust. Further, the colour of lighting is also an important factor, with findings suggesting it is important for lights to be as close as possible in character to daylight, so that objects can easily be distinguished. Our streets are lined with lights to guide us home, but often they aren't just illuminating the footpath: instead, they unnecessarily extend their reach to the tops of nearby buildings and trees, which don't require their focus, while over-illuminating our eyes and leaving some people even more vulnerable.

The loss of celestial features has happened so quietly that those who do not leave their city dwellings cannot recall what they are missing out on. Thankfully, the over-illumination of our skies and dulling of our dark nights are not inevitabilities that we must accept. By lighting our environment intelligently, consciously and minimally, we can reverse the impacts of this form of pollution far more easily than any other on Earth.

Concerns about dark skies extend not just to the loss of visibility of the animals in our dark sky constellations, but also to the harm being done to animals on the ground. Light pollution is having a significant negative impact on native fauna in Australia, with artificial light leading to a reduction in the reproductive output and quality of life of various native species.

For millennia we have been able to look to the stars to guide all facets of our understanding of the world and the universe, including animal behaviour. The proper lighting of our skies and environment is imperative not just for Indigenous peoples and astronomers, but to preserve the accuracy and depth of our extensive knowledge systems and protect the holistic ecological systems within which we live. How can we look to the Dark Emu in the sky and hope for it to guide us on the emu egg cycle on the ground if artificial light is pushing an unnatural change in the terrestrial emu's reproductive lifestyle and overall collective health?

The presence of artificial white light illuminating our environment has been shown to significantly impact many native species, particularly migratory species that rely upon phenomena occurring as expected within their environments to signpost the beginning of their natural cycles. Migratory seabirds are particularly vulnerable to excessive artificial lighting along coastal regions, as the light can cause them to divert from their natural migratory routes and even to crash into illuminated human-made structures. Increased lighting also makes these birds vulnerable to abnormal instances of predation as they lose the veil of darkness they have evolved to hide within. This is inadvertently impacting the frequency with which they roost and their choices of roosting sites, and may lead to a reduction in their reproductive output. This reality is also experienced by adult marine turtles, who may be avoiding nesting on brightly lit beaches because their hatchlings are faced with disorientation and confusion when they make their vulnerable dash to the ocean.

The theme of reproductive harm is becoming all too common in investigations of the impact of light pollution on animals. The Tammar wallaby, which is native to regions of South Australia and Western Australia, is facing active harm and alterations to its reproductive habits due to artificial lighting. The presence of unnatural light is causing a shift in the timing of its breeding season, resulting in the young being born at a time of year when their natural food resources are out of season and hence low in availability. Sadly, projections indicate severely reduced population sizes in the near future.

Similarly, artificial lighting on our streets has been linked to sleep disruption in native birds such as the Australian magpie. The presence of artificial white and red light at night has been found to impact the REM sleep cycles of magpies, with notably worse outcomes experienced under the white light than the red light. White light is produced at higher temperatures than red light, and studies have found that, in comparison to red light, exposure to white or blue light at night is more likely to suppress melatonin and shift the circadian rhythm in humans and non-human species alike. Research suggests that a choice as simple as switching from white lighting to amber lighting may lessen the health impacts of light pollution on native birds. However, this is a species-dependent solution and is not enough to address the ecological harm of light pollution altogether and protect other affected species.

Another native victim of our lighting missteps is the relationship between the mountain pygmy possum and the bogong moth. The bogong moth is a migratory species that is attracted to light and serves as a main food source for the possum. It has also served as a significant feature in the diets of Aboriginal nations across south-eastern Australia, who have many different methods of consuming the once-reliable food source – from cooking them on a fire or grinding them into a flour and baking cakes to preserving and smoking them, which is said to give them an almond-like flavour. For Aboriginal people located around the Australian Alps, the migration of the bogong moth signals the time for ceremony: at one time, hundreds, even thousands, of First Peoples migrated to the Alps to participate in the annual ceremonies. The presence of artificial light in the moth's environment misdirects its natural migratory patterns, consequently reducing food resource availability for the mountain pygmy possum and harming both species' long-term survival as well as the ability of local Aboriginal nations to practise culture.

If the health impacts of light pollution on our native species are evident, then it should be natural to assume that these rules exist for human health, too, right? There is evidence that the suppression of melatonin and the subsequent shifting of our natural circadian rhythm due to blue LED lighting may contribute to certain cancers and diseases. Research has found that chronic circadian rhythm disturbance leads to a decrease in tumour suppression and an increase in breast cancer development in mice. Blue LED lighting has also been linked to photoreceptor damage in our eyes. The rods and cones are crucial components of our eyesight and colour vision with their ability to convert light into signals that are sent to the brain for processing.

Another source of light pollution particularly relevant to Kamilaroi skies is that of gas flares erupting from natural-gas plants. Gas flares are an integral part of the operational and safety management systems for gas plants, meaning that wherever such a plant is erected, the flares are largely unavoidable. The flares do not just impact astronomical observations with the excess light they emit but are symbolic of the ecological harm that gas plants bring to their surroundings.

In Kamilaroi Country, 20 kilometres from the New South Wales town of Narrabri, a natural gas development project has been proposed. It has faced widespread protests in the past five years due to the ecological, astronomical and cultural risks it poses. The land for which the project is proposed includes the Pilliga Forests, on the sacred grounds of Kamilaroi peoples. Knowledge holder Rosie 'Bumble' Armstrong Lang told us, on the topic of the Pilliga, that 'It is important because it is where every plant exists for the Nation. It's like our Garden of Eden. If we don't protect it, we lose everything.' The forests contain close to 300 sites of significance to the Kamilaroi peoples, including camp sites, grinding grooves and ceremonial/burial grounds. The construction of the plant will lead to the destruction of parts of the Pilliga Forests, inevitably causing cultural and ecological harm.

Other core concerns about the Narrabri gas project are that it may cause significant environmental damage, particularly to local surface water and groundwater sources, could impact the health of local residents, and is in a precarious position as it is situated in an area at risk of unmanaged bushfire threats. Traditional owner and activist Polly Cutmore has relentlessly opposed the proposal since its inception, stating that 'The Pilliga forest and water is an important place for the Gomeroi people. We believe in the healing power of these waters and have historically used it for medicinal purposes. Our water is precious to us. We cannot allow further destruction and alienation of our Country and water.'

Grassroots activists suggest that investing in a renewable energy source instead of gas would mitigate the risks posed by the plant. Despite the public backlash, the project was approved in September 2020 by the New South Wales Independent Planning Commission and in late November 2020 by the federal minister for the environment, Sussan Ley.

Light pollution is not only an issue for Indigenous astronomers, impacting our stellar library and our observations as astrophysicists – it is an issue for our overall interconnected knowledge systems.

By pushing unnatural shifts in the balance of our ecosystem, we risk our knowledge, culture, health and the future of entire ecosystems.

Protecting our skies

The solution to light pollution is not to dramatically commit ourselves to the dark, but instead to invest in lighting our areas intelligently and with purpose. Any individual can take a handful of steps to properly light their surroundings, but the most meaningful change needs to be on a large scale in our public spaces.

The National Light Pollution Guidelines for Wildlife released by the Australian Government's Department of the Environment and Energy in consultation with experts, including Karlie Noon, outline six steps to follow when designing best-practice lighting for outside spaces.

They urge us to start with natural darkness. The intention is that artificially lighting a space should occur with purpose and that lighting limits should be kept in the designer's mind from the beginning. These limits would centre on how much light is necessary for that particular space, based on established standards for gauging when enough lighting has been achieved.

The next recommendation is to use adaptive or 'smart' controls for the lighting sources. These ideally enable flexible, instantaneous and remotely managed light and have the ability to switch off, dim and time the lighting. Motion sensor–activated lighting is perhaps the most obvious example that meets this recommendation.

Thirdly, the guidelines implore people to illuminate only the intended area by using directed lighting that is close to the ground. This includes the use of appropriate shielding, which in the case of a streetlight may be a cone that ensures all light emitted is being directed narrowly towards the ground without any spillage into the sky that would result in skyglow.

The guidelines then suggest that appropriate lighting is prioritised such that, within reason, the minimum amount of light needed to complete the task is all that is made available. A low-glare lighting source is strongly encouraged. People are advised to use non-reflective surfaces, and, finally, to use lights that reduce or filter out any harmful short-wavelength blue–violet light. These recommendations call for consideration at every level in a community.

The Australasian Dark Sky Alliance (ADSA), for which Krystal De Napoli is an ambassador, is a non-profit charity established in 2019 that is dedicated to educating the wider public and policy makers on the importance of dark sky conservation.

Through collaboration with local councils, astronomers, health practitioners, ecologists, tourism representatives and lighting experts, ADSA seeks to strengthen the conversation surrounding light pollution, demonstrating that the solution to minimising excessive artificial lighting in our environment will require a multi-disciplinary approach.

A society in which we emphasise the preservation of the dark skies is not an unachievable dream. It is already an everyday reality for people living under Kamilaroi skies in Coonabarabran, New South Wales – a small town at the edge of the Warrumbungle National Park. In 2016, the national park was officially designated Australia's first Dark Sky Park by the International Dark-Sky Association (IDA). The certification acknowledges that the park lies under pristine dark skies that enable the human eye to observe precious dark sky constellations and light-vulnerable features that are impossible to view from light-polluted cities.

Inside the national park is Siding Spring Observatory, a renowned optical astronomy research facility boasting many fine pieces of observational equipment, including the Anglo-Australian Telescope and the Australian National University's 2.3-metre Advanced Technology Telescope. These have contributed immensely to our understanding of dwarf galaxies, dwarf stars and optical astronomy techniques. In order to directly monitor the health of the dark skies, critical light thresholds have been placed at the observatory. The park predominantly uses lighting techniques that reduce light pollution, allowing for a fully nocturnal-friendly environment that does not harm the eating, sleeping, hunting, migrating or reproductive tendencies of its inhabitants. The certification has served as motivation for Coonabarabran and its surroundings to follow best-practice lighting guidelines to preserve the dark skies, in a model that we hope to see adopted by the rest of the continent, and the world.

As for navigating our increasingly polluted atmosphere, many solutions are required if the communication monopolies of billionaires are to continue having free rein over our skies. Just as several companies have begun to consider mitigation tactics to avoid the increase in skyglow, all companies must be responsible for their contribution to an already polluted space. Given the near-misses that have already taken place and the estimated 20 000 pieces of space debris already floating in our atmosphere, orbital pollution reduction is required. Further, reducing planetary pollution – whether in Earth's orbit or on land – will produce better outcomes for all of humanity. If we are giving these companies access to these spaces, why shouldn't they clean them up? In a seminal 1995 book, *Orbital Debris: A Technical Assessment*, the US National Research Council described four ways in which space debris can be moved from Earth's cluttered orbit: forcing objects out of orbit with ground-based instruments such as lasers, reducing the lifetime of objects by accelerating their natural decaying process, moving objects to areas with less pollution, or actively removing them from orbit. A minimally polluted atmosphere is possible with the right approach and know-how and, of course, money, of which these companies have plenty.

However, the issue of space junk is not just about removing it all – which couldn't be done even if we wanted to. Some pieces of space junk, like Vanguard 1, the oldest human-made object in orbit, have historical significance. In her recent book *Dr Space Junk vs the Universe*, Dr Alice Gorman highlights the many obstacles to the removal of space junk and says that decisions on what is to be removed should be made 'from an informed position. We need to know which objects do have cultural significance in orbit, from local, national and global perspectives. And we need to understand how their changing orbits may relate to collision risk.' She further concludes: 'It will be a while before we see large-scale space debris removal. We should use this time to plan a cultural heritage management strategy that will be both effective and practical.'

With all of these light pollution mitigation options, there is one more crucial step that must be taken by projects that aim to protect our environment. By having Indigenous people – not just Australian Indigenous people but Indigenous people from around the world – at the table and seeing them as stakeholders in the future of space exploration, sustainable living and the world more broadly, we can secure a future for all, not just for the corporations.

✱ *Galaxy in the desert*, p. **49**
   Space cowboys, p. **183**

GAPS IN THE RESEARCH

Jane McCredie

Back in 2008, psychologist Jeffrey Arnett wrote a much-cited article criticising American psychology research for focusing almost entirely on just 5 per cent of the human population (Americans) while claiming to reveal generalised truths about the whole of humanity.

His examination of research published in six leading psychology journals had found around 70 per cent of authors and their sample populations were American, with almost all the remainder coming from other English-speaking countries or Europe.

The studied population was even narrower in one journal, where two-thirds of participants in American studies were undergraduate psychology students, prompting Dr Arnett to suggest a more accurate title for the publication might be *Journal of the Personality and Social Psychology of American Undergraduate Introductory Psychology Students*.

Enrolling students is a cheap and easy way to recruit subjects, but it's hardly representative even of American society, let alone a broader range.

'No other science proceeds with such a narrow range of study,' Dr Arnett wrote of his findings.

It is difficult to imagine that biologists, for example, would study a highly unusual 5 per cent of the world's crocodile population and assume the features of that 5 per cent to be universal. It is even more difficult to imagine that such biologists would be aware that the other 95 per cent of the world's crocodile population was vastly different from the 5 per cent under study, and highly diverse in habitat, eating habits, mating practices, and everyday behaviour, yet show little or no interest in studying that 95 per cent and continue to study the 5 per cent exhaustively while making universal claims.

Humans are not crocodiles, but there's plenty of evidence a failure to include diverse populations in research does not lead to the best health care for those outside the fence.

We know, for example, that when dermatology algorithms are based on research conducted predominantly in pale-skinned people they can fail those with darker skin.

The historical tendency to study predominantly male populations has also led to poorer outcomes for women in conditions ranging from cancer, to cardiometabolic disease, to mental illness, as multiple studies show.

And, in psychology, it's been suggested clinicians who brought American-style therapeutic interventions to Sri Lanka in the wake of the 2004 tsunami, without adequate consideration of local factors, may have done more harm than good.

In 2010, a group of Canadian researchers looked at comparative psychological data to examine just how valid generalisations made on the basis of research in limited populations might be. Not very, they concluded, coining the acronym WEIRD (Western, Educated, Industrialised, Rich and Democratic) to describe the most-studied group.

In domains ranging from visual perception, to analytic reasoning, to fairness and cooperation, they found members of WEIRD societies were 'among the least representative populations one could find for generalizing about humans'. In fact, they were often outliers.

'We need to be less cavalier in addressing questions of *human* nature on the basis of data drawn from this particularly thin, and rather unusual, slice of humanity,' they suggested.

Not much changed over the following decade, according to lead author Professor Joseph Henrich, despite the paper having been cited almost 10000 times.

'On one level, I feel like there's a lot more enthusiasm around addressing sample variability,' he told *UNDARK* magazine. 'But if you actually look at the numbers, the latest numbers coming in the last few years don't actually show any shift in the diversity of samples.'

A recent follow-up to Dr Arnett's 2008 paper, co-authored by him and looking at the same journals, found there had been some change. Over the intervening years, American representation had dropped to just over 60 per cent of authors and samples, though the gap was largely filled by studies from other Western nations.

The proportion of the world's population now represented in the journals may have risen from 5 per cent to 11 per cent as a result, but the majority of humanity is still being ignored, the authors write.

Improving representation of the other 89% is an 'ethical, intellectual and professional imperative', they argue, given the influence these journals wield in current practice and future directions in psychology.

There's nothing wrong with studies conducted in specific populations, even American psychology undergraduates.

But if we want to make grand claims about the whole of humanity, we'd better make sure they're included in the process.

✳   *A whole body mystery*, p. **113**
   Why facts have ruffled feathers in the birding world, p. **238**

DO WE UNDERSTAND THE BRAIN YET?

Elizabeth Finkel

In August 2013 the cover of *Cosmos* magazine – *Decode your brain* – suggested we were on the verge of a new era of understanding how the brain works. As we approach the finish line on yet another 'decade of the brain' – one in which multi-billion-dollar campaigns have sought to reveal the workings of the most complex thing in the universe – it seems a good time to ask: are we there yet?

It wasn't the first decade of the brain. Previous iterations had promised a similar neural booty, such as new treatments for schizophrenia or spurring the development of AI. But this campaign was on steroids. For one thing, it was a twin effort.

In early 2013, Europe launched its flagship 'Human brain project', the goal of which was to simulate a human brain inside a computer within a decade. (Spoiler alert: it didn't happen.) The US countered with President Obama's Brain Research Through Advancing Innovative Neurotechnologies (BRAIN) Initiative, tasked with more pragmatic aims, such as developing circuit diagrams of the brain.

Overall, neuroscientists were armed with techniques that even two decades earlier had been the stuff of sci-fi.

Think microscopes that peer into the brain of a living animal to record signals from individual brain cells or neurons. Add the ability to switch particular brain circuits on or off with pinpoint precision using a pulse of light, a technique called optogenetics. Mix in brain atlases with the dynamic resolution of Google maps, from full relief technicolour maps of the wrinkled brain surface down to microscopic scale charts of the brain's cellular structure.

Add circuit diagrams dubbed 'connectomes' to satisfy the most exacting electrical engineer. Just as geneticists needed the genome – the complete genetic code – to understand the logic of life, so too the connectome would underpin the logic of brain function.

This amazing bag of tricks has enabled researchers to begin the task of linking the electrical signals in brain circuits to such elusive things as behaviour. 'The last decade has seen us move towards closing the explanatory gap,' says Australian National University neuroscientist Professor John Bekkers, who has spent his career analysing brain circuitry.

So: do we finally understand the brain?

Known knowns

It's a meme that the brain tends to be understood in terms of the most advanced technology of the day.

For imperial-era Romans, the brain was an aqueduct. The 17th-century natural philosopher René Descartes saw it as a hydraulic machine like the ones that moved statues in Versailles. Among late-19th-century folk it was comparable to a telephone exchange. The 20th century finally nailed it: the brain is a computer. It takes in information, stores and processes it and delivers an output; moreover, it does so by sending electrical signals through its circuits.

Of course it doesn't have the architecture of 20th-century computers. It's more akin to what the 21st has delivered – the neural networks that recognise faces in smartphones, and have now given us chatbots that pass the Turing test. It's no surprise that brain and bots share similarities in their architecture – these machines were modelled on our brain architecture in the first place.

Still, artificial neural networks are a crude facsimile of a human brain, whose 80 billion neurons and 100 trillion connections give rise to our perceptions, intelligence, emotions and consciousness.

'It's the complexity of scale,' notes Professor Gerry Rubin, director of Janelia Research Campus near Washington, DC, where the imaging techniques to spy on individual brain cells were developed and the fruit fly connectome project began.

Moreover, he says the brain computer may not be logical. 'It was built by evolution, not an engineer. The analogy I like is you went from the Ford model T into a Maserati without ever being able to turn the engine off.'

Brains are also famously more efficient at learning than artificial neural networks are – something that Dr Brett Kagan and his colleagues at Cortical Labs in Melbourne are exploring in 'dish brain' – human brain cells that can learn to play *Pong* – a first generation computer game.

So yes the brain is fantastically complex and different from the computer on your desk. But if the goal is to decode brain signals, we've made some fantastical strides – at least, if the last decade of headlines are anything to go by.

Neuroscientist Professor Jack Gallant at the University of California, Berkeley, can tell what movie a person is watching by decoding their brain waves on a functional magnetic resonance imaging (fMRI) machine. Dr Joseph Makin at the University of California, San Francisco, inserted electrodes into the brains of epileptic patients undergoing diagnostic tests and converted their thoughts to text. Professor Doris Tsao, also at UC Berkeley, can identify the face a monkey is looking at by reading signals directly from wires connected to 205 brain cells.

Then there are the mice who can be triggered to hunt and kill at the command of a laser light – by optogenetics. In another study, the romantic urges of voles were triggered by using optogenetics to activate a different set of circuits.

These examples seem to proclaim loud and clear: we *are* learning how to crack brain codes.

Yet according to neuroscientist Professor Karel Svoboda, at the Allen Institute in Seattle, these are 'parlour tricks'.

Tsao agrees: 'What we know about the brain, including my own work, is trivial.'

Known unknowns

No doubt, these brain scientists are reprising the age-old truism articulated by explorers since the time of Aristotle the more you *know*, the more you know you *don't* know. Or as Tsao puts it, 'whatever you already understand is trivial and boring'.

Nevertheless, as neuroscientists peel back the curtain on old mysteries, new vistas open up – both down into the molecular details and up into the vast cloud of emergent properties.

For Svoboda, the direction of interest points deeper into the underlying circuitry.

When he refers to mind-reading breakthroughs as 'parlour tricks', he means they are relying on correlation. Researchers like Gallant and Makin record brain signals when a person is watching a movie or articulating words, then feed them into a machine-learning algorithm that learns to associate the patterns, in much the same way these algorithms learn to detect cats on your phone.

For Svoboda – who aims for nothing less than reverse engineering the circuitry of the brain – that's not very informative. He and his colleagues are making headway in reverse engineering the mystery of short-term memory: the type that allows you to remember 10 digits long enough to tap them into your phone or follow instructions to turn right or left at the next street.

The mystery of long-term memory, that needed to permanently remember the phone number, was at least partially revealed in the 1960s. When neurons in the hippocampus (a seahorse-shaped structure involved in making memories) fire together in synchrony with neurons from other parts of the brain – say five times a second for several seconds – that solders them together in a circuit. That soldering was dubbed 'long-term potentiation'. In 2014 Roberto Malinow, a neuroscientist at UC San Diego, artificially induced that soldering by rapidly flicking an optogenetic light switch on and off to create a fake memory. The experiment proved that long-term potentiation was the mechanism behind long-term memory.

Short-term memory has remained more elusive. One long-standing theory proposed by physicist Professor John Hopfield in the 1980s suggested it involved reverberating signals between a set of neurons – as if they were humming a tune.

Svoboda has now provided evidence that this actually takes place. A recent experiment in his lab conducted by Kayvon Daie found that a circuit of some 50 neurons in the anterior lateral motor cortex (a part of the brain known to make decisions about movement) held the memory of whether a mouse should lick left or right to get a drink of water. Daie pre-trained the mice with a musical tone. A high pitch meant a drink of water lay to the right; a low pitch on the left. He observed what was happening in the mouse's brain via a tiny window in its skull that had a microscope attached to it, with a view of about 500 cells. Thanks to some nifty genetic engineering, every time a neuron fired a signal, it flashed a fluorescent light detected by the microscope. (Calcium is released when a neuron fires and the neurons were fitted with a calcium-sensitive dye.) Just prior to each lick, some 50 neurons fired together for a few seconds with a particular frequency, like a recognisable hum. Could they be encoding the memory of licking right? To check, Daie artificially stimulated those same neurons with a light switch. The mice licked right.

Away from the mice, Daie retreats to his computer to test models that describe how these networks behave. One of the most promising is the 'attractor model', which theorises that these circuits establish pre-existing templates to help complete a memory. This may be linked to our minds' strong tendency to complete patterns or see a face in a cloud.

Another stunning case of reverse engineering behaviour comes from fruit flies. Insects have legendary navigational abilities: bees unerringly find their way from the hive back to a food source and communicate its whereabouts to others via their waggle dance; foraging desert ants navigate across hundreds of metres of featureless landscape back to their nest.

Fruit flies aren't quite in the same league, but they have the basic navigational kit. Along with colleagues at Rockefeller University, Professor Larry Abbott – a physicist-turned-neuroscientist based at Columbia University – has decoded it. Fruit flies held in place by miniature harnesses roamed a virtual environment while a microscope attached to their heads recorded the activity of individual brain cells.

It turns out that to keep track of where they were, fruit flies carried out a mathematical calculation taught to high school students: vector addition. The ability to reverse engineer the circuitry of a navigating fly relied on having the entire wiring diagram, the connectome.

'In flies we can now test theories with the precision I was used to in physics,' enthuses Abbott.

Fly brains are a very long way from ours, but that doesn't mean they won't hold compelling lessons for decoding the human brain: evolution tends to re-use good inventions. A dramatic example is the eyeless gene, first discovered because it was crucial for the development of the fruit fly eye. It was subsequently found to be crucial to human eye development too.

For Professor Stephen Smith at the Allen Institute, the compelling questions are even more fine-grained. Smith has spent much of his career focused on the synapse, the place where connections between neurons are strengthened or weakened. Like sprawling tentacles, a single neuron is equipped with thousands of incoming synapses, each relaying a hopeful message from another neuron. Whether or not a neuron will accept that invitation to join part of a brain circuit is determined by what happens at the synapse. And that, believes Smith, is determined by the genes in play there. It turns out that neurons use more genes than any other type of cell. Hundreds of chemical messengers called neuropeptides are deployed at the synapse, different ones in each of the 4000 or so neuronal cell types that Smith has analysed. It is these neuropeptides, he believes, that determine which neurons will link into circuits, like those that underlie short-term memory.

Unknown unknowns

Other researchers are champing at the bit to leap to higher dimensions.

Tsao's work to date has reverse engineered how the monkey brain reads faces. She found it takes only 205 neurons in a region of the brain called the inferotemporal cortex to encode a face. The neurons are arranged in six face patches. Cells in each patch are tuned to a different facial feature. Some act like rulers to measure the distance between the eyes; others detect the face's orientation – is it looking left or right? Yet others are tuned to the colour of the eyes or hair. In a process reminiscent of the way a detective assembles an identikit, it is the combined information delivered by these face patch cells that lead the monkey, or Tsao, to identify a specific face, regardless of its orientation. The coding logic seems to generalise to other functions of the IT cortex, such as identifying whether an object is animate (like a cat) or inanimate (like a box).

'It's a beautiful example of encoding, neuron by neuron,' says Abbott. 'A few years back, I'd have said nobody's ever going to get there.'

Tsao feels less triumphant. 'I don't think any principle we understand is interesting yet,' she says. 'Everything we understand really well is still "feed forward", wiring more and more complex feature detectors. I think there has to be something more to the brain than that.' A feed forward pathway transmits information forward without incorporating feedback en route, like kicking a football towards the goal. By contrast, the goalkeeper who assesses the crosswind and ball spin is employing negative feedback.

For Tsao the next goal is to figure out how the brain incorporates various forms of feedback and puts it all together – what she refers to as the 'outer loop' of the brain computer. How, for instance, does the brain bind perceptions together in three-dimensional space to give us a model of the world? Tsao suspects that the posterior parietal cortex (a region lying roughly at the top of the primate brain towards the back) will hold clues.

Professor Ethan Scott at the University of Melbourne may beat her to it – in a juvenile zebrafish. Scott can watch this transparent fish's entire brain thinking – a symphony of 100 000 flashing neurons. He can also record the particular sections that perform when the fish listens to threatening sounds, while also sensing the flow of water and maintaining its balance. These signals all appear to be bound together in an area of its brain called the tectum – the equivalent of a mammal's superior colliculus. Like many 21st-century neuroscientists, Scott has his work cut out trying to decode the activity of thousands of neurons firing at once. 'The problem used to be collecting data like these,' he says. 'Now we are flooded with data.' Progress, he says, will rely on 'collaborations with theoreticians and mathematicians'.

Perhaps some of the outer loop properties – like consciousness – will always defy any attempt at reverse engineering.

Once a system gets to a certain level of complexity, it may be beyond the scope of reverse engineering. In a famous 2017 example, neuroscientists Eric Jonas and Konrad Kording tested whether their approach to reverse engineering brain circuits would allow them to explain the workings of a much simpler system – the 6502 gaming chip for playing *Donkey Kong*. They couldn't.

And no one has any idea how deep learning machine algorithms eventually arrive at their answers.

For Abbott that's not surprising. Coming from physics, he's comfortable with the idea that the ability to describe matter changes with scale. A hydrogen atom is completely describable by quantum mechanics, but that description can't be used for a plank of wood. 'In neuroscience I think we could understand what absolutely every neuron in the brain is doing and we still won't have an understanding of something like consciousness,' he says.

For Tsao, 'The path to understanding consciousness is going to come from AI. I think until we can experiment with it like the way we can with vision, we won't understand it.' And with the current performance of AIs, she thinks the day they develop consciousness is close.

So do we understand the brain yet?

Despite the breakthroughs of the last decade, most neuroscientists say we're just at the very beginning. It's as if we've discovered an alien computer. We've unpicked some of its hardware and are just learning to decode some of its simpler routines, but the mysterious outer loops loom before us, like a vast impenetrable cloud.

Whether we can ever penetrate that fog remains an open question.

The researchers interviewed here, equipped with ever-more-fantastical instruments, are accelerating their efforts to push ahead at all scales: drilling down to the microcircuitry of single synapses like Steven Smith; decoding discrete circuits, like Karel Svoboda and Larry Abbott; computing the thinking brain of zebrafish like Ethan Scott; and seeking to explain how the primate brain binds perceptions, like Doris Tsao.

Each would offer a different timeline for finally 'understanding the brain.' But there's one thing they all agree on. The next decade – with its ever-accelerating dialogue between artificial and natural intelligences – will be the one to watch.

✳ *Trials of the heart*, p. **34**
The psychedelic remedy for chronic pain, p. **195**
Isolation, p. **226**

BATS LIVE WITH DOZENS OF NASTY VIRUSES – CAN STUDYING THEM HELP STOP PANDEMICS?

Smriti Mallapaty

Randy Foo fills a pipette with orange juice from a bottle clearly labelled 'For Bats ONLY'. He and his colleague Rommel Yroy are seated at a biological safety cabinet, wrapped in blue gowns and wearing face shields, gloves, scrub pants and shoe covers. Peeping out of Yroy's clasped hands are two round, glossy black eyes, two slight, pointy ears and the furry snout of a young, male cave nectar bat (*Eonycteris spelaea*). It is wriggling and squealing, occasionally protruding its long, pink tongue to lick driblets of the sweet drink. This is a little treat for enduring a transfer in a blue cotton bag from its cage to the laboratory, followed by a quick weigh-in and inspection for injuries along its stretched-out wings and dense, fur coat. 'The younger guys are generally a little bit more feisty,' says Foo.

The bat in Yroy's hand is one of some 140 cave nectar bats housed in a research breeding colony in Singapore – the first in Asia. Foo, who manages the colony and is affiliated with Duke–National University of Singapore (Duke–NUS) Medical School, and Yroy, a veterinary technician at SingHealth Experimental Medicine Centre, have nurtured the bats for years. The original 19 members of the colony were caught using butterfly nets under highways around Singapore in 2015 and 2016; the first pups arrived a couple of years later.

The colony was set up by Lin-fa Wang, a virologist at Duke–

NUS Medical School to create a controlled setting for studying bat biology, including the inner workings of their immune system.

For Wang, who has spent decades studying bats and infectious diseases, the colony has been a research boon, allowing him to ask questions about, for example, the cells that make up bat immune systems and how they respond to an infection. Now that the bats are breeding productively, the team's research can be replicated more easily. They have shared bat tissue with about a dozen teams around the world. 'Bats have become a hot topic,' says Wang.

Wang's research niche has become more crowded since the emergence of SARS-CoV-2. Attendance at talks and conferences about bats is rising – at one symposium hosted last year in the United States, there were 30 per cent more participants compared with the same event organised before the pandemic – and funders are ploughing money into studies of bats and infectious diseases: in 2021, for instance, both China and the United States announced specific funding pots for research into bats and viruses.

Of particular interest is the bat immune system, especially its ability to tolerate viruses that are deadly to people and other mammals – from Ebola to Nipah and severe acute respiratory syndrome (SARS). Although bat immunity is poorly understood, its consequences are clear: bats are thought to be the source of various catastrophic viral outbreaks in humans.

The field is now at an inflection point, fuelled partly by the pandemic. Researchers who have spent decades studying infections in bats, together with enthusiastic newcomers, are developing and applying new tools to the question of how bats can live with such dangerous pathogens. Some hope that those insights could one day lead to treatments for tackling infections in people and ways to prevent viruses spilling over from bats.

'There are going to be some huge steps forward in the next two or three years in terms of bat virology and bat immunology,' says Tony Schountz, an immunologist at Colorado State University in Fort Collins.

Mangoes and melons

The bats in Wang's colony are a precious resource, and the researchers treat them accordingly. Over the years, they have tweaked the bats' diet and environment to keep the animals healthy. The bats enjoy fresh chopped melon, papaya and mango, powdered milk and a sweet-smelling, nectar-like liquid. A burlap sack hangs from the top of each cage to give the bats – who live in groups of about 25 – some privacy and darkness. Foo plans to introduce further enrichments when the colony moves to a facility with larger cages later this year.

'The colony has given us everything we wanted,' says Wang. His office is adorned with souvenirs collected over the years – a 'batman' keychain, a bat-printed mug, resin-encased bat specimens, framed drawings of bats.

The researchers have studied the bats' genomes and the diversity of viruses the animals host. They have also used bats' cells to develop airway organoids – mini organs grown from stem cells. Their current experiments are focused on the bats' response to infection, their ageing process and their very active metabolism during flight. The colony is a valuable research resource.

But studying bats and their viral tenants has been a hard grind, because the tools available to researchers are quite limited.

Colonies are expensive to establish and maintain; bats have longer pregnancies and fewer pups than the standard laboratory mouse. Only a handful of the more than 1450 species of bat have been bred in research colonies around the world. These include Wang's cave nectar bats, Jamaican fruit bats (*Artibeus jamaicensis*) in Fort Collins, Colorado, Egyptian fruit bats (*Rousettus aegyptiacus*) on the island of Riems in Germany and big brown bats (*Eptesicus fuscus*) in Hamilton, Canada. Harder to find are colonies featuring the key coronavirus-hosting horseshoe bats (*Rhinolophus* spp.). 'People have tried and failed' to breed those bats, probably because researchers don't know enough about their roosting preferences, says Aaron Irving, an infectious diseases researcher at Zhejiang University in Haining, China.

Trapping wild bats comes with its own logistical and safety challenges, and bat cells are notoriously difficult to propagate in cell culture.

Some elements of the toolkit used for other lab animals have been lacking for bats. There is a dearth of monoclonal antibodies, which immunologists use to tag immune cells and proteins. For a long time, there was no high-quality genome for a bat species. Immunologists working with mice and human tissue have been 'completely spoiled' by access to these tools, says cell biologist Thomas Zwaka at the Icahn School of Medicine at Mount Sinai, New York City.

The lack of tools means that researchers still don't have a clear picture of the 'basic architecture of the bat immune system', says Peng Zhou, an infectious diseases researcher at the Guangzhou Laboratory, China.

But years of work by established researchers and an influx of newcomers are yielding new tools and methods, including high-quality genomes and lab-made bat tissue. The next decade will see exciting insights, says Emma Teeling, a bat biologist at University College Dublin, Ireland. 'And the only reason we're going to be able to do this is due to this new generation of tools.'

There is more money for bat work too, and more research papers. References to bats in immunology articles have more than tripled, from about 400 in 2018 to 1500 in 2021, according to the scholarly database Dimensions. A start-up has raised US$100 million in venture capital funding, hoping to use knowledge gained from bat research to develop therapies for conditions from cancer to inflammation and ageing.

The latest research is filling in details of the biological mechanisms underpinning the bat immune response, including the identification of cell types that are potentially unique to bats. Researchers are also unravelling the diverse ways in which bat species tolerate viral infections. This can help determine whether there is 'one global mechanism that applies to all of the bats, and all of the viruses', says Diane Bimczok, a mucosal immunologist at Montana State University in Bozeman, who has jumped on the bat bandwagon. 'We don't know if that's the case.'

The flood of interest in this previously niche field has been somewhat overwhelming, says Hannah Frank, an evolutionary ecologist at Tulane University in New Orleans, Louisiana, who received a grant to study bat immunology from a fund created for this field by the US National Institutes of Health (NIH). 'I'm really excited about where we are,' says Frank. But the more researchers dig into the bat immune response across species, the murkier their current conclusions will get, she says. 'We're also going to realise just how complex this question is.'

Viral vessels

Researchers find a lot to puzzle over in bats. They are 'super cool animals', says Zwaka. They are the only mammals to have evolved flight, and they use sound waves to locate objects in the dark. They live exceptionally long lives for their small size, and have a low incidence of cancer.

But the trait that has brought bats into the spotlight in recent decades is their ability to host a rich collection of viruses. Certain species, especially horseshoe bats, accommodate an exceptional diversity of coronaviruses, which include those closely related to SARS-CoV-2. Some species also host viruses such as rabies, Ebola and Marburg. Bat genomes are studded with viral remnants.

What is it about bats that allows them to tolerate viruses without showing signs of infection? Over the years, researchers have come up with some theories, which the new tools could help to refine.

Studies reveal that some bat species mount a robust first defence against invaders. Even in the absence of a foreign threat, some species maintain high levels of interferons – molecules that raise the alarm and ramp up efforts to disable viruses – which could allow the animals to quash viral replication quickly. Bats also have an expanded repertoire of genes encoding proteins that interfere with viral replication or stop viruses from leaving cells. Their cells are equipped with an efficient system for disposing of damaged cell components, known as autophagy, which has been shown to help clear viruses from human cells.

When pathogens do intrude, bats typically don't overreact with an outsized inflammatory response, which is often responsible for much of the damage caused from an infection. Bats have several ways to tame the inflammatory response, such as suppressing the activity of large multiprotein molecules known as inflammasomes. Instead of spending huge amounts of energy getting rid of a virus completely, they seem to tolerate low levels of its presence, says Irving. 'There's kind of a peace treaty' between bats and the pathogens they host, says Joshua Hayward, a virologist at the Burnet Institute in Melbourne, Australia.

Both established researchers and newcomers to the field are now beginning to look beyond the swift and indiscriminate part of the bat's defence – the innate immune response – and towards the slower, more targeted adaptive response that retains information about a pathogen and springs into action when it meets its foe again. Adaptive immunity is restricted to a few specific cell types and a 'pain in the butt' to study, says Frank.

Some researchers are also studying the link between bats' immune response and their ecology to better understand when and where they shed viruses, and the risk of spillover to other animals. This work could illuminate the environmental factors that put bats under stress, and whether that increases shedding and spillover risk.

Into the bat cave

Some of these questions took Javier Juste to an abandoned dam in Cádiz, Spain, one night in May 2020. He crept into a concrete tunnel at the dam and collected two horseshoe bats from their roost.

The coronavirus pandemic was swirling – and scientists knew that the virus had probably originated in bats. Juste's colleagues in the United States were keen to get their hands on bat tissue, hoping that they could grow the cells and use them to explore how deadly viruses can jump from bats to people.

Bats that can host coronaviruses are not commonly found in North America, but they are all over Europe. So Juste, who is based at the Doñana Biological Station in Seville, Spain, had agreed to net and send bats from a roost near Cádiz to New York City – in the middle of one of the strictest Covid-19 lockdowns in the world, and the mass grounding of international flights.

Speeding down deserted highways with two bats, Juste and a colleague reached the airport in Madrid the next morning. Outside the FedEx warehouse, working in the boot of their car, the researchers euthanised the animals, sliced and squeezed the bones and organs into six specimen tubes, and stashed the tubes in a coolbox to keep the cells alive. They hauled their precious cargo to the counter, with minutes to spare before the plane door closed. 'It was probably one of the longest days of my life,' says Juste, who had spent months acquiring permits for the journey that day.

Some 26 hours later, the samples arrived at Zwaka's lab. The lab was nearly lifeless, and the package hadn't fared much better; many of the cells had already died. Zwaka, who had never handled bat tissue before, rushed to extract marrow from the wing bones, and cut squares of skin from the diaphanous, rubbery wings.

His team used the specimens to produce stem cells – a commonplace tool for studying biology and disease in other species but difficult in bats. These induced pluripotent stem (iPS) cells, described in a paper in *Cell* in February, have already revealed some intriguing insights about the close evolutionary ties between bats and viruses. Zwaka says that the cells have led his team 'down quite a rabbit hole in terms of biology'.

Together with Teeling and other colleagues, Zwaka sequenced the RNA expressed by these cells and found an abundance of sections that were essentially viral fragments, many of which were originally coronavirus genomes. The viral gene expression was higher and more diverse in pluripotent cells than in both bat skin cells and in pluripotent cells from mice and humans. What's more, the pluripotent bat cells actually used the viral fragments to make what appeared to be virus-like particles.

'The results were extraordinary,' says Teeling. When you make bat stem cells, you essentially 'wake up all the fossilised viruses that you find in the genomes'. The cells seem to systematically suck up viral information in their genomes – 'almost like a sponge' – and then express it. This makes these bat cells an environment conducive to viruses, says Zwaka. But what exactly this means for how these bats have learned to coexist with viruses isn't yet clear. One possibility is that the genetic inserts somehow protect the bats from the negative outcomes of a viral infection, just as a vaccine would, says Teeling.

The researchers now plan to use the stem cells to generate lung, gut and blood tissue, as well as infecting the cells with viruses. Zwaka hopes to use the tissue to better understand bat immunity, and eventually 'to develop strategies for human health'. Other researchers are using bat organoids developed using stem cells extracted directly from bats to answer similar questions.

1000 genomes

Cells and tissues are one thing, but one of the most important resources for cell biologists is a genome. Before 2020, there were about a dozen bat genomes, of varying quality. That year, Teeling and her colleagues described the first high-quality genomes for six bat species, each belonging to a different genus and each with its protein-coding genes clearly tagged.

The project was part of a global genome consortium, called Bat1K, co-founded by Teeling and aiming to create high-quality genomes for every bat species. A surge of interest and funding since the pandemic, including from biotech companies, has resulted in the sequencing of some 80 bat genomes so far, says Teeling.

The availability of high-quality genomes has transformed the bat immunology field. It has facilitated large-scale studies of RNA molecules and proteins and provided a way to classify immune cells, to some extent overcoming the lack of monoclonal antibodies. The genomes will be the 'foundation for many, many studies,' says Marcel Müller, a virologist at the Charité University Hospital in Berlin.

Irving is working with the Bat1K consortium to expand its collection of Chinese horseshoe bats (*Rhinolophus sinicus*), which are the hosts for the closest known relatives of SARS-CoV-2. They have sequenced ten new genomes – including four from the horseshoe, or rhinolophid family. In a preprint posted in February, Irving, Teeling and colleagues found that bat genomes had many more genes involved in immunity and metabolism under positive selection than do other mammals. They took a closer look at one gene, *ISG15*, which expresses an antiviral protein that plays an important part in hyperinflammation observed during SARS-CoV-2 infections in people.

In cell experiments, they found that the rhinolophid and hipposiderid versions of the protein lacked an amino acid found in most other mammals. The change seemed to keep it from leaving cells, and probably prevented the protein from triggering an inflammatory response. Such proteins could hold important clues for how bats live with deadly viruses and could inspire therapies in people, says Irving.

Among the trendiest techniques enabled by high-quality genomes is single-cell RNA sequencing, in which researchers take a cell of interest and analyse its RNA contents to explore the cell's components and how they work.

Last November, Wang's team published the results of its first single-cell sequencing foray. The researchers infected cave nectar bats with pteropine orthoreovirus, a virus commonly found in this species but which doesn't make them sick. In the bats' lung cells, the researchers identified the fingerprints of many familiar immune cells, including T cells, as well as some unfamiliar cells.

At present, most single-cell studies in bats are simply catalogues of immune-cell activity. In unpublished work, Schountz and his colleagues infected Jamaican fruit bats with H18N11 influenza A virus and looked at which cells it targeted. They found that the virus targets macrophages – immune cells that patrol the body and gobble up pathogens – which has not been seen before for influenza A viruses. The single-cell study gave the researchers a great head start for more detailed cell-culture experiments. 'At the very least, it gives you some ideas of where you should start looking,' says Schountz.

Other scientists are using RNA sequencing to compare bat and human cells. For instance, Nolwenn Jouvenet, a virologist at the Pasteur Institute in Paris, and a new entrant to the field, is combining this technique with CRISPR gene editing in cell lines from a range of bat species to look for differences in the innate immune response of bat and human cells. Ultimately, Jouvenet hopes to identify the genes responsible for controlling viral replication.

For some questions, only a whole bat will do. At his colony, Schountz wanted to test whether fruit bats, which are not natural hosts of SARS-CoV-2, can be made susceptible to it. So his team used a viral vector to express the ACE2 receptor, which SARS-CoV-2 uses to enter cells, in the bats' lungs, and then infected the bats with SARS-CoV-2. They found that the bats produce T-helper cells specific for the virus; these cells are key players in the adaptive, targeted immune response. Stimulating the T-helper cells produced small proteins known to regulate inflammation, which could explain the tempered inflammatory response in bats. The results were posted as a preprint in February. Schountz is planning a new bat facility, with construction starting in June and to be completed by 2024, with flight rooms to house larger bats such as flying foxes (*Pteropus* spp.). Other teams also plan to establish colonies, including one for Jamaican fruit bats at Montana State University.

Back in Singapore, it is hot and humid under the bright midday Sun, but inside the vivarium, the temperature is noticeably cooler. Yroy and Foo are speaking loudly above the noisy hum of a machine outside.

The bats seem unfazed, huddled together, upside-down, in a dark corner of their cage, letting out occasional squeaks. 'They are used to people coming and walking around the cage,' says Foo. Earlier that morning, Yroy had laid new plastic sheets to catch their droppings, and hung fresh bowls of water. Soon it will be feeding time. 'So far, they are quite happy,' says Foo.

✱ *Long Covid: After-effect hits up to 400 000 Australians*, p. **154**
A mystery of mysteries, p. **203**

TAWNY FROGMOUTHS

Anne Casey

Bucking plummeting numbers,
clustered at dusk –
braced against a stiff westerly
on a teeming branch
of the jacaranda –
two tiny tight-lipped
shut-eyed fluff-bundles
huddle up to their mother,
with her one yellow eye
primed, silver-grey plumage
simmering against
the snakeskin bark, a solitary
male back-to-back
with his brood
bringing up the rear

as we cluster with our sons,
goggle-eyed beneath
purple bells flying,
in stunned gratitude –
reluctant to budge
lest we break the spell.

✱   *A mystery of mysteries*, p. **203**
    Why facts have ruffled feathers in the birding world, p. **238**

ONBOARD THE SPACE STATION AT THE END OF THE WORLD

Jackson Ryan

On Christmas Day 2021, all was quiet aboard the RSV *Nuyina*. The vessel was holding position in Storm Bay, at the southern end of Australia, only a few kilometres from shore. The crew were holding their tongues.

Every so often, the ship's announcement system would *bing bong* and the shipmaster's voice would boom out a warning. Don't flush the toilet. Don't bang any doors. Keep conversation to a minimum. The vessel's acoustic scientists had dropped a listening device off the back of the *Nuyina* into the ocean early on Christmas morning to test how quiet the ship was as it sailed. Any banging or flushing – even farting, joked the shipmaster – would mess with the experiment.

Three days earlier, I'd trudged up the slightly swaying gangplank of the *Nuyina* (pronounced 'noy-yee-nah'), the Australian Antarctic Division's new A\$529 million (US\$373 million) icebreaker, with a backpack full of electronics, 10 woollen sweaters, seven paperback books, five notepads, two pairs of shorts, two Android phones and one Polaroid camera, joining 66 expeditioners and crew for the ship's maiden voyage to Antarctica.

The bizarre Christmas was an unofficial start to the multiyear process of commissioning the ship's scientific instruments. 'This voyage is a really critical first step in setting up the next 30 years,' said Lloyd Symons, the voyage leader.

Though Antarctica rests isolated over the planet's southern pole, it's not protected from the disastrous effects of human-induced climate change. Some parts of the continent are warming faster than anywhere else on the planet, leading to changes in sea ice conditions, weather patterns and wildlife populations. And the effects are not localised; what happens in Antarctica does not necessarily stay in Antarctica. Heating in the region will have global impacts, as sea levels rise and ocean circ ulation is disrupted.

The *Nuyina* is a ship perfectly positioned to monitor and understand how these changes will affect the rest of the world. It's been described as 'Disneyland for scientists', but given that it's dotted with antennas and huge cranes and lined with specialised scientific instruments across its hull, it feels much more NASA than House of Mouse.

It's essentially the International Space Station, a rocket ship and a luxury cruise liner all in one, and its destination is just as isolated and hostile to human life as low Earth orbit. Living on the ship and learning its rhythms feels about as close as you can get to being in space without leaving the planet.

I spent 39 days on the space station–cruise ship as it sailed to two Australian Antarctic outposts and a monstrous glacier in December 2021 and January 2022. I interviewed the scientists and engineers on board to understand its capability to perform cutting-edge research at the bottom of the world and how its design and instruments will be used to assess the impacts of climate change in the future. And yes, I spent time in the ship's theatre watching John Carpenter's *The Thing*.

Over the journey, the ship would need to tick off a number of milestones, commissioning pioneering scientific instruments as it sailed the Southern Ocean. It would also need to accomplish a series of challenging firsts, including refuelling Australia's lifeline to the Antarctic, the Casey Station outpost, with 1 million litres of a special diesel blend. None of the *Nuyina*'s systems had been used in the Antarctic before – it would be a journey into the unknown.

But before the ship had even left port, it encountered problems with its alarm system, delaying departure by two days. When the *Nuyina* first encountered sea ice, an instrument used to map the ocean floor snapped off the underside of the ship and was lost to the Southern Ocean.

Worse still, as the vessel sailed into the Antarctic Circle and approached the icy continent, it appeared as if the *Nuyina* was destined to avoid the ice altogether.

Into the ice

You've probably never heard of a drop keel, but on the *Nuyina*'s maiden voyage, the component threatened to sink the ambitions of the science team.

Shaped like an aircraft wing and covered in sensitive instruments, the drop keel juts down out of the underside of the ship like a hangnail. It's designed to be lowered during sailing to conduct surveys, enabling researchers to map the ocean floor and see and hear the marine life in the Southern Ocean.

During the first week of the voyage, after lowering the starboard drop keel about 75 centimetres below the hull, engineers encountered a problem: the keel wouldn't retract. A pin designed to hold it in place was stuck.

If the *Nuyina* was to smash through ice, there was a danger the crushed remnants would slip below the waves and slam into the drop keel. The chief concern was for the sensitive scientific instruments that adorn the keel. Jagged fragments of ice could do real damage.

No matter what the crew attempted, a quick fix seemed futile. At one point, an expeditioner returning to their cabin noted a ship engineer sitting cross-legged on the floor, schematics laid haphazardly all around. Other expeditioners wanted to help fix the problem themselves. One volunteered to don scuba gear, enter the freezing water and swim underneath the *Nuyina* to manually pluck out the pin.

That plan was never given any real thought.

The vessel's major mission on its first voyage was to reach Casey Station, on the East Antarctic fringe, and deliver roughly 1 million litres of fuel. Before that, it would need to resupply Australia's Davis station further west, using helicopters to cart supplies, such as food, booze and mail, from ship to shore.

On the approach to Davis, about ten days into the journey, the ship was confronted by an ocean dense with heavy sea ice. Thousands of floes jutted across the surface of the ocean at acute angles, as far as the eye could see. With the drop keel down and vulnerable, the shipmaster and crew had to decide whether they'd push through and risk damage, or skirt the boundary of the ice.

They chose the latter, keeping to the edge of the sea ice before snaking through lighter floes and into the bay near Davis Station. But the manoeuvering posed another question: would Australia's new icebreaker actually *break* ice on its first voyage to Antarctica?

That question would linger until refuelling at Casey Station was complete. In the meantime, there was science to do.

Standing 198 centimetres and confidently rocking an exposed scalp, Rob King faces a persistent challenge aboard the *Nuyina*: making sure his head doesn't get split open by a door frame.

When I find him leaning over one of the *Nuyina*'s purpose-built aquariums full of prized Antarctic crustaceans and other miniature sea beasts, it feels like a mythical scene: a Titan surveying his domain.

King, a krill biologist with the Australian Antarctic Division, has been studying the Antarctic's most important food source for over three decades. In that time, he's developed techniques to improve catches in the Southern Ocean and helped design world-class aquariums at the division's headquarters in Kingston, Tasmania. He's on board the *Nuyina* to try to capture a fresh stock of krill from the Antarctic.

Antarctic krill, 5-centimetre-long crustaceans with bulging black eyes and a transparent exoskeleton, are the Southern Ocean's keystone species. They support the Antarctic ecosystem by providing food for wildlife such as whales, seals and penguins, and they're incredibly abundant, making up about one-fortieth of all animal biomass on Earth – comparable to the biomass of humans.

Typically, capturing krill involves trawling. Researchers drop a net off the back of a ship, which balloons out in the ocean around a swarm of krill. Closing the net traps the crustaceans and pulls them back up to the surface. But the *Nuyina* includes a world first: as the ship was being constructed, engineers added three holes into the hull that connect to a room dubbed the 'wet well'.

It's a room King dreamed up some 15 years ago to catch Antarctic krill in pristine condition and bring them back to Tasmania.

As the *Nuyina* sails south, water flows in through the holes, which are connected to the wet well, or what King describes as 'effectively a very wet room'. As the water moves into the ship, it brings marine life with it – krill, copepods, phytoplankton – as if vacuuming them up from the ocean.

'It's a way of collecting specimens without dragging a big net through the water,' King says.

On the *Nuyina*'s maiden voyage, the wet well would finally be put to the test. King's dream would be realised – or not.

In the early morning of 30 December 2021, King and *Nuyina* aquarist Anton Rocconi opened the wet well's valve for the first time at sea, sampling the Southern Ocean's wintry water.

The water passed through pipes in the starboard side of the ship, across a raised table in the center of the wet well and into a tank at its end. Almost as soon as it gurgled on, sea life started pouring in.

Sea butterflies, wings still flapping gracefully, began to fill buckets. Amphipods, voracious shrimplike predators, followed. Krill were barrelling in too, but not the Antarctic species the pair were hoping to capture. On that first operation, the *Nuyina* was still 1600 kilometres or so from the Antarctic krill's stomping grounds. The prize catch of the voyage was yet to be seen, but the run buoyed spirits.

The pair became more optimistic because the catch was in such pristine condition. Trawl nets can do damage to sea life by crushing creatures as they're ripped from the ocean, but the wet well is a far more gentle hand. At lunch, the high-energy Rocconi, curly mullet swinging side to side behind his head, seemed more certain than ever the wet well would work.

He and King started waking up early, at around 3 am, a time when krill come close to the surface to feed. They'd switch on the wet well and try again.

On 2 January, the pair struck a rich vein of crustacean gold. The first Antarctic krill sloshed through the pipe and across the filter table, slipping by Rocconi. The moment was the culmination of decades of work for King, but in a weird twist of fate, it was data technician Tess Chapman who claimed the honour of spotting the *Nuyina*'s first Antarctic krill.

In the early days of the voyage, Rocconi said he was hopeful of catching 500 critters by the time the *Nuyina* returned to Tasmania. The day after the first catch, on 3 January, the team caught 2000. Hooting and hollering broke out in the room as water splashed across the filter table and underfoot. The wet well worked, and it worked superbly.

'That was what I dreamed it could always deliver,' King said.

Through the ice

On the *Nuyina*'s science deck there's a hole. It sits in the corner of a room, and the deep sea instrument that hangs over it smells like decaying fish.

If you fell down the shaft, which is just under 14 metres long, you'd end up wet, cold – maybe even dead.

The hole is the *Nuyina*'s 'moon pool,' and it allows scientists to access the Southern Ocean through the middle of the ship. You can think of it like the moon door from *Game of Thrones*: an opening to the great beyond below. Fortunately, the *Nuyina*'s moon pool has two hatches, at the top and bottom, so expeditioners can't be kicked into it unwillingly (at least, not from what I saw).

It's a critical component of the *Nuyina* because it allows scientists to access the ocean *underneath* the ice. Typically, research vessels deploy instruments or underwater drones over the side or back of the ship, lowering them slowly to the ocean floor with kilometres-long cables. But when the ship is amid ice in the Antarctic Ocean, that's impossible. Instead, if you want to see what's lurking underneath, you have to get off the ship and cut a hole through before you can lower instruments down.

The *Nuyina*'s moon pool sidesteps that problem, giving scientists access to a world that was previously unreachable. It's a place where alien sea creatures swarm and where cold ocean water rests before circulating around the planet. It's a place that has barely been explored at all – but it could be critical for understanding climate change.

Climate scientists worry about how global warming is affecting Antarctica's abyssal depths. Recent studies show the water lying at the bottom of the Southern Ocean is getting warmer and less salty, which could spell disaster for global ocean currents and how they circulate, throwing weather systems into chaos. But the effects of the change can be evaluated only with constant monitoring. With a moon pool, the *Nuyina* can deliver constant access, sampling the icy world and understanding how it changes.

On its maiden voyage, *Nuyina* sent down instruments through the moon pool while anchored in the Southern Ocean, sampling water and taking temperatures and conductivity measurements. Watching instruments descend down the shaft feels like seeing spacecraft slowly launched from the ISS, departing into a murky black void. And future experiments will bring this imagery even closer to fruition as expeditioners and scientists send down autonomous underwater vehicles and drones.

In March, scientists on a South African icebreaker in Antarctica's Weddell Sea showed how useful underwater drones can be in surveilling the world beneath the ice: they discovered the wreck of the *Endurance*, Antarctic explorer Ernest Shackleton's ship. The 106-year-old wreck was full of life. Ghostly sea squirts, spindly sea stars and even a pale white crab had colonised the sunken ship.

Drones will be key to understanding the ecosystems that exist below the East Antarctic ice and will help reveal more about these unknown worlds – and others across the solar system. NASA, for instance, has tested autonomous drones that could reveal the alien worlds that exist under the ice surrounding Antarctica. These tests will pave the way for drones to be sent to the frozen moons of Jupiter and Saturn, dropping into subsurface oceans on Europa and Enceladus.

Data on ice

Johnathan Kool is hunting for the elusive and mysterious Planet Nine, a theoretical cosmic body lurking beyond the orbit of Neptune.

The observation deck of the *Nuyina* might seem like a strange place to search for planets, but this isn't real life – it's a card game known as The Crew, which works a bit like Hearts. Whenever he has a morning off, Kool joins me and the ship's two doctors to resume the game and our planetary search.

In the ship's newsletter, Kool, the data centre manager at the Australian Antarctic Division, gets compared with the History Channel's 'Aliens' meme guy. But he's far less dishevelled and much more honest than the likeness suggests.

For most of his career, he's been working with unwieldy, massive datasets, running complicated computer simulations and models that take into account hundreds, if not thousands, of variables. 'I think I was big data before it was "big data",' he often remarks.

His major task on the *Nuyina*'s maiden voyage is to ensure that the ship's science instruments work in harmony with its data systems, delivering real-time information to anyone working aboard the ship. He says that if he's doing his job well, no one should really notice him, and his major challenge in wrangling data follows on directly from his love of games like The Crew.

'I've always been attracted to the cooperative ones,' he says. 'I like trying to win against the system.'

And on the *Nuyina*'s first voyage, the system is a handful. The ship records information from 70 instruments, including sensors that measure particles in the water; sonar that looks for schools of fish and krill; meteorological instruments measuring ultraviolet radiation, humidity and air temperature; webcams; CCTV; hydrophones; echo sounders and CTDs (which measure conductivity, temperature and depth). Kool, and data technician Tess Chapman, have to ensure the endless stream of ones and zeros is filtering in correctly and in a way that's comprehensible to the ship's scientists.

In an ideal world, those scientists wouldn't even need to be on the *Nuyina*, Kool says. Instead, they'd be able to access data collected by the ship, in real time, from anywhere in the world. A biologist in Portugal could follow blue whales as they feed on krill; a meteorologist could monitor changes in air pressure to help predict weather conditions.

On the first voyage, this would've been impossible. The *Nuyina*'s unstable satellite internet connection was one of the biggest bugbears for expeditioners. It had a maximum download speed of 4Mbps, the average internet speed in a US household in 2008, and an upload speed of 1Mbps. Downloading a PDF on the ship would take an entire morning. Calling home via WhatsApp was an impossibility. Sometimes, there'd be no internet connection at all.

It's easy to see this as a First World problem, but improved connectivity is key to opening up science in the Antarctic. Big data is the most valuable thing to researchers, and in the modern world the internet forms the backbone of those systems. Kool even mentions Starlink, the SpaceX satellite broadband network, as a good example of potentially cheap ways to improve connectivity in the future.

Kool isn't focused just on data, though. His other job is to supervise the AAD's seabed mapping program. The *Nuyina* is an invaluable new ally in a worldwide project to map the entire ocean floor by 2030, using acoustic instruments to illuminate the world below.

During its maiden voyage, the *Nuyina* showcased just how instrumental it'll be in uncovering the secrets of the Southern Ocean floor.

Under the ice

Open Google Maps and zoom out until you can see the Earth as an orb, floating in space. Focus on the ocean, and you'll clearly see the shadows of ridges and valleys, scars on the planet's face that crisscross the ocean floor.

The study of these curves, peaks and valleys on the seafloor is known as 'bathymetry', and there's evidence for humans performing these studies as early as 3000 years ago, in ancient Egypt. Much of what you see on Google Maps is determined by satellites. But these aren't actual observations. They're rough estimates of how the planet looks beneath the waves, obtained by studying data from those satellites.

The true contours of the ocean floor largely remain a mystery.

'We know more about the surface of Mars than we do our own planet,' Kool says. 'We have about eighty per cent to ninety per cent of Mars mapped, whereas the oceans are only at about twenty per cent.'

Bathymetry is fundamental to understanding the geological processes that affect our planet and the history of the Antarctic. Tens of thousands of years ago, when sea levels were lower, glaciers extended further out from the continent, leaving deep voids in the Earth that are now covered by water.

'There's a physical story that's preserved on the bottom of the seafloor,' says Matt King, a professor of polar geodesy at the University of Tasmania and director of the Australian Centre for Excellence in Antarctic Science. 'Human eyes have never seen that.'

Floyd Howard, an acoustics officer aboard the ship, explains that the vessel contains active acoustic instruments known as echo sounders, which emit sound and listen for the echo to bounce back off the seafloor. 'This is what a bat does at night, or a dolphin does in the sea,' he says.

During the first voyage, Howard and acoustic officers Jill Brouwer and Alison Herbert used the ship's echo sounders to map a 2.2-kilometre-deep canyon extending underneath the Vanderford glacier in East Antarctica. In 2018, NASA scientists revealed that Vanderford and nearby glaciers had been losing almost 30 centimetres of ice each year since 2009. They reasoned this change might be due to warmer ocean currents from the north winding their way to the Antarctic, sneaking up on the glaciers and melting them from below.

The *Nuyina*'s echo sounders were also able to see underneath the glacier for over 3 kilometres, showcasing the ship's ability to image parts of the Antarctic never seen before. The videos the instruments produce feel otherworldly – rainbow-coloured trenches show the ship's path through the unmapped dark. A spacecraft charting an abyss.

By providing such a detailed map of what lies beneath, the *Nuyina* forms a critical part of predicting Vanderford's future. 'It's a glacier that is retreating and sensitive to a warming climate, but it turns out we don't have a lot of observations of this region,' notes Felicity McCormack, a glaciologist at Monash University in Melbourne who's studying Vanderford.

She says the data collected by the *Nuyina* will be incredibly important when it comes to looking at how the glacier might change in the future.

Vanderford was the penultimate stop on the *Nuyina*'s journey before it made for home. The voyage was missing one critical milestone: breaking the ice.

Break the ice

Petersen Bank is an iceberg graveyard.

Its shallow waters trap mammoth icebergs until they melt or, shrinking, free themselves and drift out to sea. But the *Nuyina* comes to rest in front of a seemingly endless expanse of 'fast ice', water frozen to the continent's shore. From the crow's nest, a few icebergs are visible, as big as mountain ranges and poking out of a truly alien landscape, a white sheet that extends to the horizon.

With station resupply and refuelling complete – and despite the drop keel remaining exposed underneath the ship – a decision is made: the *Nuyina* will break Antarctic ice for the first time.

After idling the ship at the edge of the ice, the master sends the command to steam straight ahead. The ship's engines power its thrust forward, taking the first nibbles at the edge of the fast ice and causing Adélie penguins to scatter and dive into the water.

Further afield are Emperor penguins, the largest species, but they appear unperturbed.

The cracking of ice barely registers over the sound of the wind buffeting the ship, but if you get close enough, you can hear the *Nuyina* rumbling forward. You can hear the ice give way, unleashing shrill cracks as it slips below the bow. Then, if you head to the back of the ship, you can see the trail of destruction that lies in her wake – literally.

Smashing through the frozen mass caused a chunk of ice, bigger than a football field, to begin drifting out to sea. A jagged line runs backward, to where the *Nuyina* entered the ice. Bits of ice have accumulated in the newly opened passage, bridging the gap. An Adélie penguin, chased by a seal, zooms out of the water and onto the ice bridge.

Expeditioners fill the upper decks of the ship to snap their keepsake pictures with telephoto lenses and smartphones. This is really the definitive moment for the *Nuyina* – breaking ice in Antarctica is what the ship was built to do. The ship rests within the ice for hours. Surrounded by the purest white as far as the eye can see. Very few people will ever experience a moment like this and, with that realisation, I'm reminded of space again. A vast nothingness fundamental to understanding who we are and why we're here.

Eventually, the master orders the *Nuyina* to back up. The wildlife flees, the expeditioners return to cabins and workstations, and the ship sets course for her final destination: home.

In just four weeks, blessed by the weather and oceans, the *Nuyina* completed a number of historic firsts. It was able to supply Casey Station with 1 million litres of fuel and deliver helicopters and supplies to Davis. Its scientific instruments were put through their paces, mapping features of the ocean floor, capturing krill in perfect condition and reaching just above the seabed to assess the ship's state-of-the-art moon pool. Finally, it had broken ice.

The ship sailed into Hobart on 30 January 2022, as the sun crested over Mount Wellington.

Seventeen days later, refueled and restocked, the *Nuyina* was once again blasting her way to Antarctica.

❋   *Buried treasure*, p. 7
    A subantarctic sentinel, p. 85

ANTIMATTER: HOW THE WORLD'S MOST EXPENSIVE – AND EXPLOSIVE – SUBSTANCE IS MADE

Carl Smith

It's the most expensive substance on Earth, costing quadrillions of dollars for a single gram. It's also likely the most explosive substance on the planet.

Michael Doser – who works in the only factory making it – describes this reaction as 'probably the most violent process you can think of because the full mass of the object disappears and transforms into energy'.

And based on what we know about this terrifying-sounding substance, the Universe probably shouldn't exist at all. So what is it?

Antimatter.

It doesn't sound like it should be real, but 'it does exist', says Professor Doser, a physicist who studies the properties of antimatter at CERN, the European Council for Nuclear Research. This international scientific institution in Switzerland is home to the Large Hadron Collider, and it regularly exposes the hidden particles that make up our universe.

Lesser known is its role in studying the anti-particles of the Universe.

Professor Doser leads a team studying this strange, expensive, explosive stuff in the wonderfully named Antimatter Factory. Here they create and capture this bizarre anti-stuff.

What is antimatter? You can think of it as matter's evil twin.

Professor Doser actually thinks matter might be the evil half of this equation, with antimatter being the 'good guy'. But the point is: antimatter is the opposite of matter. It's exactly the same as matter, except all the electrical charges of its component parts are reversed. This is why it's so explosive. When a bit of matter comes into contact with its evil antimatter twin, they cancel each other out, releasing all the energy stored inside them.

'[When] a proton and antiproton annihilate each other, their mass completely disappears,' Professor Doser says. 'So this is by far the most energetic process that you can think of.'

By converting all their mass into energy, you're getting more bang for your buck with an antimatter explosion.

'In the case of a chemical reaction, you're transforming only about a millionth of the mass of the object of the molecule into energy,' Professor Doser says.

The violence of an antimatter reaction was clearly demonstrated when a tiny pinch of the stuff exploded over Vatican City ... in the fictional Dan Brown epic, *Angels and Demons*. Thankfully, outside the realms of science fiction, we won't see antimatter destroying cities anytime soon.

'Even in that hypothesis you'd still need a gram of antimatter, which would take 10 billion years to accumulate,' Professor Doser says.

And for my Trekkie friends out there, that also means fusion-powered warp drives like those on Star Trek ships are unlikely to be a thing any time soon.

So how is antimatter made?

To create antimatter you just need to create matter. Simple? Nope. Expensive? You bet.

The recipe they use at CERN's Antimatter Factory to achieve this feat is:

1   Take a proton (a charged subatomic particle).
2   Speed it up enormously.
3   Crash it into an iridium block.

One in every million collisions creates a proton-antiproton pair.

The basic principle is that so much energy is concentrated at a single point that it creates mass – the mass of matter. And yes, bizarrely, energy can become the mass of matter – and vice versa. This equivalency is most famously described in Einstein's equation:

e (energy) = m (the mass of matter) × c (the speed of light) squared

But whenever this happens – when loads of energy gets concentrated and turned into the mass of matter – antimatter is born too. 'Antimatter appears every single time matter appears,' Professor Doser says.

The cost of creating antimatter like this makes it the world's most expensive substance. Professor Doser once estimated how much it would cost to make antimatter in large amounts. 'One 100th of a nanogram [of antimatter] costs as much as one kilogram of gold,' he says. After a bit of number crunching that means a gram of antiproton antimatter would cost an absurd 5 quadrillion euros. That's 5 thousand trillion euros.

There's not really any point translating that to Australian dollars because it's absurd either way.

Other sources of antimatter

And yet, a piece of fruit makes antimatter too. And so do we!

'Bananas are a perfect unit for antimatter production. It's one antiparticle per hour, approximately,' Professor Doser says.

As radioisotopes in bananas decay, they release pairs of

electrons and anti-electrons. The same process happens in the human body too, so we're all creating anti-electrons. But to understand the properties of this mysterious anti-stuff, apparently anti-electrons won't cut it. Professor Doser and his colleagues need antiprotons.

'You need 2000 times more energy to make [anti-protons],' he says. 'So we actually need infrastructure like at CERN, accelerators that will produce enough energy locally in a very small spot to produce pairs of an antiproton and a proton.'

But ... why would we bother?

Ah yes, the multi-million dollar question.

There are a few answers.

The first is that the technology developed in CERN's Antimatter Factory has been applied in medical imaging tools called PET scanners.

The second is that CERN is interested in fundamental research – understanding things without knowing how this knowledge could be applied.

And the final one is that it might help us solve a fairly enormous cosmic conundrum: why the material universe exists.

The Universe probably shouldn't exist

At the moment of the Big Bang, all the energy of the Universe was concentrated and exploded.

'We actually expect that the whole Universe – since there was lots of energy around at the moment of the Big Bang – should consist of equal amounts of matter and antimatter,' Professor Doser says. 'The big surprise is that it doesn't.'

There is no antimatter left in the Universe from the Big Bang that we're aware of, he says.

Which is fortunate. If the Big Bang led to equal parts matter and antimatter forming, these probably would have then bumped into each other, obliterated one another, and then presumably exploded again.

'We want to study it to see why it's not here anymore and why the Universe isn't just empty.'

So, what's their working theory as to why our evil antimatter twins didn't just cancel everything out, long ago?

'The best explanation that we have found up to now is to say that there's a slight difference in the properties of particles and antiparticles,' Professor Doser says. This means that although equal amounts of matter and antimatter should have formed, they weren't quite equal, he adds. 'One particle is left over out of a billion, and this one particle out of a billion is everything we see in the Universe. All the galaxies, the clusters of galaxies, the stars, the planets, us. 'We're the leftovers in this model.'

It's a pretty convenient explanation. But it's not the only one.

He says an alternative hypothesis is that we're living in a part of the Universe filled with matter – but other parts might be full of antimatter. In other words, antimatter planets, antimatter stars or antimatter galaxies could be a thing.

'If we don't find a difference between matter and antimatter, then that's going to be the only remaining explanation,' Professor Doser says.

Unravelling this cosmic conundrum is what the researchers at CERN's Antimatter Factory are trying to do. But so far, this mysterious anti-stuff remains elusive. The team hasn't found any other meaningful differences between matter and antimatter.

And if you're concerned that this work doesn't warrant fooling around with such a violently explosive anti-substance, Professor Doser says there's no need to worry.

'We make such minute quantities that even if you were to destroy all the antimatter that we're making in the course of a year, it wouldn't be even enough to boil a cup of tea.'

✱ *Galaxy in the desert*, p. **49**
A universe seen by Webb, p. **78**

CONTRIBUTORS

PAUL BIEGLER is a journalist, academic, and former doctor specialising in emergency medicine. His science writing has been published in *The Age*, the *Sydney Morning Herald*, *Good Weekend*, the *Australian Financial Review* and *Cosmos Magazine*, and he is the author of *The Ethical Treatment of Depression*, which won the Australian Museum Eureka Prize for Research in Ethics. His new book *Why Does It Still Hurt? How the Power of Knowledge Can Overcome Chronic Pain* is published by Scribe.

JACINTA BOWLER is an Adelaide-based science journalist, fact checker and audio producer who has written about everything from quantum laptops to superbugs. They have penned pieces for some of Australia's most respected media organisations and were published in *The Best Australian Science Writing 2022*.

TABITHA CARVAN is senior science writer for the Australian National University and a freelance writer on the side. Her work has been featured in publications including the *Guardian*, the *Sydney Morning Herald* and *The Age*, *The Saturday Paper*, *Crikey*, *Junkee*, *Australian Geographic* and *The Best Australian Science Writing 2022*. She is also the author of the memoir *This Is Not a Book about Benedict Cumberbatch* (Fourth Estate).

ANNE CASEY, originally from Ireland, is a Sydney-based poet/writer and author of five poetry collections. A journalist for 30 years, her work is widely published internationally, ranking in the Irish Times Most Read. She has won literary awards in the US, UK, Ireland, Australia, Canada, India and Hong Kong, most recently American Writers Review 2021 and the Henry Lawson

Prize 2022. She is a doctoral researcher in archival poetics at UTS. Find her at anne-casey.com and @1annecasey

JO CHANDLER is a journalist and journalism educator. Her focus is on deep, explanatory reporting across a diverse range of topics. She has earned numerous distinctions including the UNSW Press Bragg Prize for Science Writing (2012 and 2014). Her work has featured in the *New York Times*, the *Guardian*, the *Atlantic*, the *Monthly*, *Griffith Review*, *Good Weekend*, *Cosmos Magazine* and *New Scientist*, among others. She is a senior lecturer at the Centre for Advancing Journalism, University of Melbourne.

ANGUS DALTON completed a cadetship with the *Sydney Morning Herald* in 2022 and is now a science reporter for the paper. He is the co-founding editor of *Sweaty City*, a magazine about climate change and urban ecology, and his writing has appeared in *The Best Australian Science Writing 2019* and *2022*, *Australian Geographic*, *Overland* and *Kill Your Darlings*.

KRYSTAL DE NAPOLI is a Gomeroi award-winning author, astrophysicist and science communicator devoted to the advocacy of Indigenous knowledges and equity in STEM. Krystal is co-author of *Astronomy: Sky Country* (2022), winner of the Victorian Premier's Literary Award People's Choice Award (2023), shortlisted for both the Victorian Premier's Literary Award for Indigenous Writing (2023) and the *Age* Book of the Year (2022). Krystal is the host of the weekly radio show *Indigenuity* on Triple R 102.7FM.

ELIZABETH FINKEL is a biochemist who switched to journalism. She co-founded *Cosmos Magazine*, serving as editor-in-chief from 2013 to 2018 and is now editor at large. She is the author of *Stem Cells*, which won the Queensland Premier's Literary Award and *The Genome Generation*. Besides journalism, she serves as a vice-chancellor's fellow at La Trobe University and on advisory committees for La Trobe University Press and Zoos Victoria.

LAUREN FUGE is an Adelaide-based writer, currently undertaking a PhD on how we tell stories about climate change. She was awarded the 2022 UNSW Press Bragg Prize for Science Writing for a piece she wrote for *Cosmos Magazine*, where she also works as an editor. Her debut nonfiction book, *Voyagers: Our Journey into the Anthropocene*, will be published in 2024.

REBECCA GIGGS is the author of *Fathoms: The World in the Whale*, a work of nature writing. In 2023 she is a literature fellow at Akademie Schloss Solitude in Stuttgart, Germany.

ALICE GORMAN is a space archaeologist and author of the award-winning book *Dr Space Junk vs the Universe: Archaeology and the Future* (NewSouth, 2019). She is an associate professor at Flinders University in Adelaide and a heritage consultant with over 25 years' experience working with Indigenous communities in Australia. In 2021, asteroid 551014 Gorman was named after her in recognition of her work in establishing space archaeology as a field.

AMALYAH HART is a freelance science journalist raised in the UK and based in Melbourne. With a background in archaeology, she's obsessed with all things science: her regular beat is climate and energy, but she covers the gamut from human evolution to robotics.

ZOE KEAN is a multi-award-winning science writer with a focus on evolution, ecology and the environment. You can find her words in the *Guardian* (UK and Australia), the ABC, *The Best Australian Science Writing 2022*, *Cosmos Magazine*, the *Tasmanian Inquirer* and the BBC. She is currently a features reporter and radio producer for ABC Radio Hobart. Her love for science cannot be contained to the page – she also gives talks, makes science TikTok videos and regularly appears on live radio.

ALICE KLEIN has been the Australia reporter for *New Scientist* magazine since 2016. She has a PhD in chemistry and was previously a reporter at *Australian Doctor* magazine. She mainly writes about health and the environment.

SMRITI MALLAPATY is a reporter with *Nature*, and covers the Asia-Pacific region. She has a master of science degree in environmental technology from Imperial College London, and reports on subjects including biological and environmental sciences, space, publishing and community in Asia, and Covid-19.

JANE MCCREDIE is an award-winning writer, journalist, critic and public speaker. She is the founder and director of the Quantum Words Festival of writing about science, held in Sydney and Perth. Her writing about science, medicine and the environment has been widely published in Australia and internationally. Jane is the author of *Making Girls and Boys: Inside the Science of Sex*, and is a former editor of *The Best Australian Science Writing*. She is a previous senior judge of the NSW Premier's Literary Awards and has been the CEO of Writing NSW for the last ten years.

FIONA MCMILLAN-WEBSTER is a science writer based in Brisbane. She has a BSc in physics and a PhD in biophysics, and writes about everything from botany and ancient chemistry to quantum physics and modern biomaterials. Her writing has appeared in *Cosmos Magazine*, *Australian Geographic*, *National Geographic*, *Forbes* and other publications. Her first book, *The Age of Seeds: How Plants Hacked Time and Why Our Future Depends on It*, was recently published by Thames & Hudson Australia.

BIANCA NOGRADY is an award-winning freelance science journalist, author, and founding president of the Science Journalists Association of Australia. She writes about science, health, and the environment for outlets including the *Guardian*, *The Saturday Paper*, *Nature*, *WIRED*, the *BMJ* and *MIT Technology Review*. Bianca is the author of books including *Climate Change: How We Can Get to Carbon Zero*, and *The End: The Human Experience of Death*.

KARLIE ALINTA NOON is a Gamilaroi astrophysicist with over a decade's worth of experience in science communication. She has extensive knowledge in Indigenous heritage and STEM education and is passionate about making STEM, including Indigenous knowledges and values, accessible to everyone. Karlie is co-author of the award-winning book *Astronomy: Sky Country* and is currently undertaking a PhD in astrophysics at the Australian National University.

MEREDI ORTEGA is an Australian poet and writer living in Scotland. Her recent poems have appeared in the *Times Literary Supplement*, the *Poetry Review*, *Meanjin*, *Magma*, *Gutter* and *Poetry News*. She contributed to the deep mapping anthology *Four Rivers, Deep Maps* (UWAP).

MIKI PERKINS is a multi-award-winning senior writer, with 16 years of reporting experience as a journalist at *The Age*. For the past three years she has been *The Age*'s specialist writer on climate change and the environment, covering a broad range of topics including the rise of climate action in the courts, the renewable energy transition and the decline of biodiversity in the natural world.

NICKY PHILLIPS is a science journalist based in Sydney, Australia. For the past six years, she has been a news editor and the chief of the Asia-Pacific bureau for *Nature* magazine. Before joining *Nature*, Nicky worked as a science reporter for the *Sydney Morning Herald* and as a radio reporter and producer for the Australian Broadcasting Corporation. She has degrees in science and journalism.

FELICITY PLUNKETT is an award-winning poet and critic. Her books are *A Kinder Sea* (UQP), *Vanishing Point* (UQP) and the chapbook *Seastrands* (Vagabond). She edited *Thirty Australian Poets* (UQP).

EUAN RITCHIE is a wildlife ecologist and conservation biologist, within Deakin University's School of Life and Environmental Sciences. He is also a councillor within the Biodiversity Council. Professor Ritchie's research team, the Applied Ecology and Conservation Research group, use ecological theory combined with extensive fieldwork and experiments to help better understand, manage and conserve species and ecosystems. In addition to undergraduate teaching, Professor Ritchie is also a prolific and passionate science communicator.

DREW ROOKE is a freelance journalist, writer, and the author of *One Last Spin: The Power and Peril of the Pokies* (Scribe, 2018) and *A Witness of Fact: The Peculiar Case of Chief Forensic Pathologist Colin Manock* (Scribe, 2022), which was shortlisted for the 2022 Ned Kelly Awards for Best True Crime Writing. He was a 2021 Walkley Our Watch Fellow.

JACKSON RYAN is an award-winning science journalist and the current science editor at CNET.com. In a previous life he studied the cells in our bones, but now he wants to write the best science stories you've ever read. He owns an extensive collection of bad

Christmas sweaters and tweets sparingly (and mostly angrily) @dctrjack

BELINDA SMITH became a science journalist after realising she wasn't going to cut it as a scientist. Based in Melbourne, she's currently science reporter at the ABC, and her work appears on the ABC News website. You can also hear her talking about science on local radio and RN. In her spare time, Bel's a GPS artist who runs maps in the shape of animals. Find her tweets @sciencebelinda and GPS art on Insta @animalpunruns

CARL SMITH is a science journalist in RN's Science Unit. He makes audio documentaries and written features. He's been an ABC News cadet, a geneticist, a reporter on *Behind the News*, a 'journalist in residence' in Heidelberg, Germany, and an animated presenter on the ABC Education series *Minibeast Heroes*. Carl also writes and co-hosts the ethics podcast for kids *Short & Curly*, and he is the vice-president of the Science Journalists Association of Australia.

HELEN SULLIVAN's features have appeared in the *London Review of Books*, the *New Yorker* online, the *New York Times* and the *Guardian*, where she is a columnist. She was a runner-up for the UNSW Press 2022 Bragg Prize for Science Writing. Her memoir, *Freak of Nature*, will be published in 2024. She is represented by the Wylie Agency in London.

HEATHER TAYLOR-JOHNSON writes on Kaurna land, near Port Adelaide. A novelist, poet, essayist and arts reviewer, she's also the editor of *Shaping the Fractured Self: Poetry of Chronic Illness and Pain*. Her latest book is a fifth poetry collection called *Alternative Hollywood Ending*. She is an adjunct research fellow at the University of Adelaide's JM Coetzee Centre for Creative Practice.

CLARE WATSON is a freelance science journalist with a background in biomedical science. Since trading her pipettes for a pen, her work has aired on ABC Radio National's *Health Report* and appeared in *Undark*, the *Guardian*, *Hakai Magazine*, *Cosmos Magazine*, and *Australian Geographic*. She writes and fact-checks for ScienceAlert, and loves finding stories that pull back the curtain on how science happens and who scientists really are. Follow @clarewhatson

SARA WEBB has always been fascinated by our very existence and the wonder of it all. Her journey to better understand the physics of the Universe has guided her towards sharing this passion with the wider world. The Universe we exist in is one full of wonder and mystery, just waiting to be explored, understood and shared. Sara's research has focused on observational transient astronomy, fuelling her fascination for the unknown.

ALAN WEEDON is a Melbourne-based photographer, and radio producer for ABC Radio National. His photographic work has been exhibited at a range of galleries including the National Portrait Gallery and the Museum of Australian Photography. Find him at alanweedon.co

ACKNOWLEDGMENTS

'Buried treasure' by Jo Chandler was originally published as
'Buried treasure: Journey into deep time' in the *Griffith
Review* on 2 August 2022.

'This magnificent wetland was barren and bone-dry. Three
years of rain brought it back to life' by Angus Dalton was
originally published under the same title in the *Sydney
Morning Herald* on 21 November 2022 and is available
at <www.smh.com.au/environment/conservation/this-
magnificent-wetland-was-barren-and-bone-dry-three-years-
of-rain-brought-it-back-to-life-20221115-p5bydw.html>

'Trials of the heart' by Nicky Phillips was originally published as
'She was convicted of killing her four children. Could a gene
mutation set her free?' in *Nature* on 9 November 2022 and is
available at <www.nature.com/articles/d41586-022-03577-
9>

'Galaxy in the desert' by Jacinta Bowler was originally published
under the same title in *Cosmos Magazine* on 24 March 2023.

'A city of islands' by Helen Sullivan was originally published
under the same title in the *London Review of Books* on
1 December 2022 and is available at <www.lrb.co.uk/the-
paper/v44/n23/helen-sullivan/diary>

'Ears' by Heather Taylor-Johnson was originally published under
the same title in *Science Write Now* in January 2023.

'A universe seen by Webb' by Sara Webb was originally published
under the same title in *Cosmos Magazine* on 28 September
2022.

'First Light' by Meredi Ortega was originally published under the
same title in *Magma* in October 2022.

'A subantarctic sentinel' by Drew Rooke was originally published as 'A subantarctic sentinel: Solving the mystery of Macquarie Island's dieback' in the *Griffith Review* on 2 August 2022.

'Model or monster?' by Amalyah Hart was originally published under the same title in *Cosmos Magazine* in June 2022.

'A whole body mystery' by Alice Klein was originally published under the same title in *New Scientist* on 28 January 2023.

'Point of view' by Lauren Fuge was originally published under the same title in *Cosmos Magazine* on 24 March 2023.

'In the shadow of the fence' by Zoe Kean was originally published under the same title on ABC Science on 9 February 2023 and is available at <www.abc.net.au/news/2023-02-09/dingo-fence-map-ecology-farming-predator-sheep-extinction/101711608>

'The Torres Strait Islander elders lawyering up to stop their homes from sinking' by Miki Perkins was originally published under the same title in *Good Weekend* on 30 July 2022 and is available at <www.theage.com.au/national/the-torres-strait-islander-elders-lawyering-up-to-stop-their-homes-from-sinking-20220609-p5asfd.html>

'Long Covid: After-effect hits up to 400 000 Australians' by Bianca Nogrady was originally published under the same title in *The Saturday Paper* on 9 July 2022 and is available at <www.thesaturdaypaper.com.au/news/health/2022/07/09/long-covid-after-effect-hits-400000-australians>

'Noiseless messengers' by Rebecca Giggs was originally published under the same title in *Emergence* on 1 July 2022 and is available at <emergencemagazine.org/essay/noiseless-messengers/>

'Space cowboys' by Alice Gorman was originally published under the same title in the *Big Issue* on 5 August 2022.

'Where giants live' by Belinda Smith and Alan Weedon was originally published under the same title on ABC Science on 24 August 2022 and is available at <www.abc.net.au/news/

science/2022-08-24/mountain-ash-trees-climate-change-fire-logging-threats/101356494>

'Talara'tingi' by Felicity Plunkett was originally published under the same title in *Plumwood Mountain* in December 2022.

'The psychedelic remedy for chronic pain' by Clare Watson was originally published under the same title in *Nature* on 28 September 2022 and is available at <www.nature.com/articles/d41586-022-02878-3>

'A mystery of mysteries' by Fiona McMillan-Webster is an excerpt from her book *The Age of Seeds*, originally published on 26 July 2022.

'"Gut-wrenching and infuriating": Why Australia is the world leader in mammal extinctions, and what to do about it' by Euan Ritchie was originally published under the same title in *The Conversation* on 19 October 2022 and is available at <theconversation.com/gut-wrenching-and-infuriating-why-australia-is-the-world-leader-in-mammal-extinctions-and-what-to-do-about-it-192173>

'Isolation' by Paul Biegler was originally published as 'Isolation: What does lockdown do to brains and genes?' in *Cosmos Magazine* on 13 June 2022.

'Why facts have ruffled feathers in the birding world' by Tabitha Carvan was originally published as 'Rooting for the antihero: Why facts have ruffled feathers in the birding world' in *Science* at ANU on 24 February 2023 and is available at <science.anu.edu.au/news-events/news/rooting-antihero-why-facts-have-ruffled-feathers-birding-world>

'Dark skies' by Karlie Noon and Krystal De Napoli is an excerpt from their book *Astronomy: Sky Country*, originally published on 26 April 2022.

'Gaps in the research' by Jane McCredie was originally published under the same title in *MJA InSight*+ on 7 March 2022.

'Do we understand the brain yet?' by Elizabeth Finkel was originally published as 'So, do we understand the brain yet?'

in *Cosmos Magazine* on 16 March 2023 and is available at
<cosmosmagazine.com/news/do-we-understand-the-brain-
yet/>

'Bats live with dozens of nasty viruses — can studying them
help stop pandemics?' by Smriti Mallapaty was originally
published under the same title in *Nature* on 21 March 2023
and is available at <www.nature.com/articles/d41586-023-
00791-x>

'Tawny frogmouths' by Anne Casey was originally published
under the same title in the *Canberra Times* on 27 August
2022.

'Onboard the space station at the end of the world' by Jackson
Ryan was originally published under the same title on
CNET.com on 2 May 2022 and is available at
'Antimatter: how the world's most expensive – and explosive
– substance is made' by Carl Smith was originally
published under the same title on ABC Science on
19 February 2023 and is available at <www.abc.net.au/
news/science/2023-02-19/antimatter-factory-physics-most-
expensive-explosive-substance/101948092>

UNSW   BRAGG   PRIZE
PRESS   FOR
LTD.   SCIENCE   WRITING

In 2012, UNSW Press launched an annual prize for the best short nonfiction piece on science written for a general audience. The UNSW Press Bragg Prize for Science Writing is named in honour of Australia's first Nobel laureates, William Henry Bragg and his son William Lawrence Bragg. The Braggs won the 1915 Nobel Prize for physics for their work on the analysis of crystal structure by means of X-rays. Both scientists led enormously productive lives and left a lasting legacy. William Henry Bragg was a firm believer in making science popular among young people, and his lectures for students were described as models of clarity and intellectual excitement.

The UNSW Press Bragg Prize is supported by the Copyright Agency Cultural Fund and UNSW Science. The winner receives a prize of $7000 and two runners-up each receive a prize of $1500.

The shortlisted entries for the 2023 prize are included in this anthology.

The UNSW Press Bragg Prize
for Science Writing 2023 Shortlist

Jo Chandler, 'Buried treasure'
Lauren Fuge, 'Point of view'
Rebecca Giggs, 'Noiseless messengers'
Amalyah Hart, 'Model or monster'
Nicky Phillips, 'Trials of the heart'
Helen Sullivan, 'A city of islands'

Winners announced in November 2023.
<unsw.press/the-unsw-press-bragg-prize-for-science-writing/>

Judges for the UNSW Press Bragg Prize 2023

Professor Merlin Crossley, University of New South Wales
Professor Elaine Baker, University of Sydney
Professor Andrew Mackintosh, Monash University
Associate Professor Caroline Ford,
University of New South Wales
Donna Lu, editor, *The Best Australian Science Writing 2023*

The successful Bragg Prize, which recognises excellence in science communication, hosts a special category for high school students. Science enthusiasts in years 7 to 10 are invited to submit an essay of up to 800 words.

A joint initiative of UNSW Press, UNSW Science and Refraction Media, with support from the Copyright Agency Cultural Fund, the prize is designed to encourage and celebrate the next generation of science writers, researchers and leaders. For an aspiring university dean of science or Walkley Award–winning journalist, this could be the first entry on their CV.

The winner receives a $500 UNSW Bookshop voucher, publication in next year's *The Best Australian Science Writing* and CSIRO's *Double Helix* magazine, and an invitation to the launch of *The Best Australian Science Writing* in Sydney. Two runners-up each receive a book voucher worth $250. The regional and city schools with the most entries receive a book pack from NewSouth Publishing, including a copy of this book. The names of the winners and details about the competition are available on the UNSW Press website: <unsw.press/the-unsw-bragg-student-prize-for-science-writing/>

The 2022 competition invited students to write a short essay on 'What do we mean by science?' Entries in 2022 were judged by a panel comprising Corey Tutt, NSW Young Australian of the Year 2020 and founder of Deadly Science; Fred Watson, Australia's first Astronomer-at-Large; Ivy Shih, editor of *The Best Australian Science Writing 2022*; Brad Thomas, science teacher, Macarthur Girls High School; and Heather Catchpole, Refraction Media, and founder, Careers with STEM.

The winning entry was 'Viral science' by Olivia Campbell, a year 9 student from Presbyterian Ladies' College, Melbourne, Victoria. Her winning essay is included on the following pages.

The UNSW Press Bragg Student Prize
for Science Writing 2022 Winner

VIRAL SCIENCE

Olivia Campbell
Year 9, Presbyterian Ladies' College,
Melbourne, Victoria

To begin: a confession. Each day, I spend two hours and 48 minutes on my phone. I derive a masochistic pleasure from doom-scrolling infographics about climate change on Instagram. When I was 10, I burned through our curtains trying to turn my grandfather's lighter into a soldering pen like I saw in 5-Minute Crafts, and it took me longer than I care to admit to realise that my sunburn couldn't be cured by massaging it with honey. I used to write poetry, and essays like this one ... but quiet contemplation is so much more lonely than letting YouTube tell me what to think.

Social media is a ubiquitous landscape of the modern world: one ruled by likes and views, overseen by some omniscient algorithm. Since the emergence of these platforms in the early 2000s, the way we communicate as a society has been entirely deconstructed and rebuilt, seemingly from scratch. And of course, when the very structure of our civilisation is so suddenly shifted, there are bound to be cracks in the foundation.

Any post on these platforms exists in an economy of clicks. Every story has to be sensational for us to bite; be reeled in. So what do we do, when the happenings of the world don't make bright enough bait? We spray-paint minnows gold.

To illustrate: in late July, an article was published in the journal *Hypertension* titled 'Association of nap frequency with hypertension or ischemic stroke supported by prospective cohort data and mendelian randomization in predominantly middle-aged European subjects'. The next day, news.com.au released a piece of their own, provocatively titled 'Do you like napping? You might also have a deadly heart condition'.

What happened here, to create such a vast chasm between these two articles – one that seems to broaden with every comment and reblog? Though this is a fairly trivial example, the real world impacts of misinformation and poor science communication are by no means inconsequential. Facebook yells that the Covid-19 vaccine is a hoax, a lie disseminated by 'Big Pharma', an experiment for which we are the lab rats – 'that will kill us'. So, thousands of people refuse to be injected ... Some of them probably die.

In the worst of times, it can seem as if science communication and modern media are two wholly incompatible beasts – one repelling the other as similarly charged poles on a magnet. But the argument here is in fact a far more nuanced one.

Social media lends a certain virality to knowledge, unlike anything we have known before. While, as discussed above, this can often lead to the widespread dissemination of fake news, it also has the potential to allow important information to rapidly reach a whole demographic of young people aged 10 to 30. Somehow, videos on subjects ranging from reproductive healthcare to the latest developments from universities find them. Simply in the process of living, children are exposed to a trove of thought-provoking concepts. These are often in the same frame of reference as dance challenges and make-up vlogs, placing them in the context of fun, relevant and dopamine-inducing content.

Then, we see the algorithm giving people exactly the type of material that will make them stay on the platform – or, in less cynical terms, the kind of science they are truly interested in, as opposed to the musty case studies in their textbooks. The micro-format of posts essentially forces them to be accessible and digestible, distilling complex ideas into an easily understandable solution, and hence providing young people an entryway into the vast world of science.

Finally, the gatekeeping of academia is abolished in the world of modern media – a benefit to society being reshaped from the ground up. Anyone can speak on a topic that matters to them. People with disabilities are able to insert themselves into the conversation about what they want and need; students can talk with their peers rather than lecture them.

All of these factors serve to cast light on STEM as an endeavour – an expedition – as opposed to a table of figures to memorise. The impact this may have on growing minds, and their desire to pursue the sciences as a career, is potentially massive. And of course, this is a field where we need diverse perspectives and experiences to draw from, both in identifying problems and developing viable solutions.

In asking what science is in the modern world, this essay has revealed that there is no true answer ... Or rather, that there are thousands. Some of these are ugly while others inspire, and far more find themselves somewhere in-between. But we cannot begrudge a field for evolving alongside its creators. Science is the basis of how we understand the universe, and this is simply what it will take for it to survive.

ADVISORY PANEL

MERLIN CROSSLEY is an enthusiastic university teacher, researcher and administrator. He trained or worked at the University of Melbourne, Oxford, Harvard and Sydney before moving to the University of New South Wales, Sydney, as dean of Science. He is now the deputy vice-chancellor (academic quality). He works on human genetic diseases and is focusing on using CRISPR-mediated gene editing to treat sickle cell anaemia. He is or has served on the boards of the Australian Museum, the Sydney Institute of Marine Science, the Australian Science Media Centre and UNSW Press, and is on the editorial boards of *The Conversation* and BioEssays.

CAROLINE FORD is a cancer researcher at the University of New South Wales with expertise in ovarian cancer, endometrial cancer and endometriosis. A long-time advocate for women in science and women's health, in 2017 she was named a Superstar of STEM by Science & Technology Australia, and in 2018 was named the Women's Agenda Emerging Female Leader in Science, Medicine and Health. An avid reader, she is also the founder of the STEMMinist Book Club, a global and virtual community focused on feminism and women in STEMM (science, technology, engineering, mathematics and medicine), which includes over 5000 members from 30 countries worldwide.

ELAINE BAKER holds the inaugural UNESCO Chair in Marine Science at the University of Sydney. She leads the university-hosted regional office of UNEP/GRID-Arendal, an official collaborating centre with the United Nations Environment Programme. She works with several United Nations organisations on international policy development primarily related to the sustainable use of resources. For over two decades she has been a global leader supporting developing coastal states and small island developing states in finalising their maritime boundaries. She is currently part of an international team driving change in the global mining industry by introducing mandatory standards for mine waste management.

ANDREW MACKINTOSH is Head of the School of Earth, Atmosphere and Environment at Monash University, and is a Chief Investigator of the Australian Research Council Special Research Initiative 'Securing Antarctica's Environmental Future'. His research aims to improve our understanding of glacier and ice sheet response to climate change, including assessing the impacts on sea level, water resources and ecosystems. He is a regular commentator in the media and was a Lead Author of the IPCC Special Report on the Oceans and Cryosphere in Changing Climate. He has a PhD from the University of Edinburgh and has held visiting positions at Columbia University and the University of Bristol. Prior to Monash, he was Director of the Antarctic Research Centre in Wellington, New Zealand.